Chromosome Maps of *Drosophila*

Volume II

Author:
Veikko Sorsa
Professor of Genetics
Department of Genetics
University of Helsinki
Helsinki, Finland

CRC Press, Inc.
Boca Raton, Florida

Library of Congress Cataloging-in-Publication Data

Sorsa, Veikko.
Chromosome maps of *Drosophila*.

Bibliography: p.
Includes indexes.
1. *Drosophila*—Genetics. 2. Chromosome mapping.
3. Insects—Genetics. I. Title.
QH470.D7S67 1988 595.77′4 87-36826
ISBN 0-8493-5857-4 (v. 1)
ISBN 0-8493-5858-2 (v. 2)

Direct all inquiries to CRC Press, Inc., 2000 Corporate Blvd., N.W., Boca Raton, Florida, 33431.

International Standard Book Number 0-8493-5857-4 (v. 1)
International Standard Book Number 0-8493-5858-2 (v. 2)

Library of Congress Card Number 87-36826
Printed in the United States

PREFACE

The natural banding of giant interphase chromosomes, due to an exact pairing of parallel chromomeres of polytenized chromatids, offers an abundance of structural landmarks necessary for exhaustive mapping of chromosomes. The occurrence of highly polytenized chromosomes in the interphase nuclei of easily preparable larval salivary gland cells has made *Drosophila* an important and preferred object for cytogenetical research. In particular, *Drosophila melanogaster,* the genome of which has also been genetically mapped in greatest detail, offers an incomparable resolution with combined molecular and chromosomal mapping of genes and other DNA sequences. For 50 years, the guide maps drawn by C. B. Bridges and the reference system proposed by him for the salivary gland chromosomes of *Drosophila melanogaster* have served the cytogeneticists working with this organism. Especially the revised versions of the Bridges' maps have been detailed enough even for the finest cytological analyses.

One limitation, quite bothersome, for instance in recent studies with immunochemical markers, is the low resolution of light microscopy. Difficulties in following and interpreting the revised reference have become more pronounced during the last 10 years. The rapid development and continuous refinement of molecular methods allowing the isolation, cloning and chromosomal localization of specific DNA sequences, have as well focused the interest on accuracy of chromosome maps.

An important but still open question is the total number of bands in the polytene chromosomes, i.e., the number of chromomeres of individual chromatids in different species of *Drosophila.* Aside from introducing new light microscopic and electron microscopic maps of the salivary gland chromosomes of *Drosophila melanogaster* compared with the revised reference maps of Bridges, another aim of this book is to assemble and present in one place the polytene chromosome maps available from the *Drosophila* species studied most extensively during recent years. In order to provide the reader with a brief overview of the historical background and our present understanding of structural and molecular aspects of polytene chromosomes, the first part of this book contains reviews on these topics.

Chapters 1 and 2 will introduce the polytene chromosomes with some historical perspectives an early studies on them. Chapters 3 to 9 are general reviews elucidating the current knowledge and hypotheses about the structural organization of polytene chromosomes. Chapters 10 to 12 mainly concern the functional organization of polytene chromosomes. Since Chapters 3 to 12 have been written in a rather concise form, their heavy contents may burden some readers, particularly those who are not aquainted with the structural details. To make for easier reading the reviews have been illustrated with micrographs and diagrams. Some of the diagrams will also conclude the most essentials from the text of the chapter.

The number of reports published in the fields studied most extensively, i.e., replication, transcription, and the chromosomal localization of genes and other sequences is so large that it has been possible to refer to only a few of them, principally those concerning the polytene chromosomes of *Drosophila melanogaster,* and then only from a structural point of view. The choice of references may reflect the author's own opinions and the evaluation of available data, his own standards. However, the aim has been to cover by way of the cited reports most of the essential information published on each field considered in the book.

Chapter 13 in Volume I and Chapters 14 to 17 in Volume II will introduce the mapping of polytene chromosomes and the use of maps in cytogenetic and phylogenetic studies. A composite list of maps available for different species of *Drosophila* (in Chapter 15 in Volume II) will help one to locate the original reports concerning mapped chromosomes or regions of them. Some examples of maps of complete genomes have been reproduced here, but in a relatively small format. The different types of maps are also illustrated by some examples.

The more complete collection of salivary gland chromosome maps published of *Drosophila melanogaster* also includes a brief historical review of mapping.

The electron microscopic division maps of the whole salivary gland chromosome complement of *Drosophila melanogaster* introduced in this book have been reconstructed from a large number of electron micrographs of thin-sectioned material produced mainly during the years 1965 to 1980. Thin section EM of squashed chromosomes started with the early attempts in the Department of Genetics, at the University of California, Berkeley and continued with further developing of methods in the Department of Genetics of the University of Helsinki. A great number of the electron micrographs used for the mapping were obtained as a ''side product'' of the localization of the *white* gene started in the Max-Planck-Institut für Biologie, at Tübingen, 1968 and including a period in the Departments of Genetics and Zoology, at the University of California, Davis, 1974 to 1975.

In this connection the author would like to express his gratitude to Professor Wolfgang Beermann from Tübingen and Professors Melvin Green and Stephen Wolfe from Davis for providing the facilities needed in this work. In Finland, the electron microscopic studies were mainly carried out in the Department of Electron Microscopy at the University of Helsinki. The preparative work, as well as the analyses of electron micrographs has been done in the Department of Genetics at the University of Helsinki from which I would like to thank the former and the present heads of the Department, Professor Emeritus Esko Suomalainen and Professor Olli Halkka for their support and interest. The studies have been financially supported by the National Research Council of Sciences of Finland.

Of the many people involved in the different phases of the EM mapping during about 2 decades, I would like to mention with special gratitude Drs. Virpi Virrankoski-Castrodeza and Anja O. Saura for their indispensable help in preparation and thin-sectioning of salivary gland chromosomes. In particular, I thank my colleagues Anja O. Saura and Tapio I. Heino for their patience in doing most of the laborious band analyses from the electron micrographs of thin-sectioned salivary gland chromosomes, which have been the basis of mapping.

The division maps were drawn at a magnification of approximately × 25,000 by using average band thicknesses estimated from electron micrographic representations of the salivary gland chromosomes made by using magnifications from about × 25,000 to about × 250,000. Besides appearing in publications, the electronmicrographs have been presented in poster exhibitions at several conferences. The maps are tentative and minor changes may be necessary, particularly, in the description of bands of puffed or otherwise difficult regions. The Bridges' doublets, interpreted now as single bands on the basis of EM analyses, may also need reexamination by direct densitometry from EM negatives and by using more stretched chromosomes.

The thickness values of bands were also used in calculating the DNA content per division and subdivision. Although still very rough and locally inaccurate, the DNA values may help geneticists approximate distances between genes and regions at the DNA level. Hopefully, the maps and tables will find use among biologists working in the fields of molecular genetics and cytogenetics, and also among those studying the phylogeny of *Drosophila*.

In writing this book I have received the most valuable help from Dr. Michael Ashburner from Cambridge (UK), to whom I am very grateful for the critical reading of the manuscript with numerous useful comments, and also for providing me with a computer list of the mapped species of *Drosophila*. Accordingly, I am grateful to Drs. David J. Finnegan and Dan L. Lindsley for their kind help in providing the most up-to-date information about-transposable elements and gene sites for the book.

The diagrams illustrating Chapter 7, Volume I and Chapters 16 and 17 in Volume II were skillfully drawn by Ms. Carin Sahlberg. Mr. Tapio I. Heino has patiently photographed all diagrams and maps, and also provided the book with the beautiful light microscopic division maps.

Last but not least I would like to thank my wife Dr. Marja Sorsa for all of her interest and support, from my earliest beginnings with the EM of polytene chromosomes up to the last lines of this book.

Veikko Sorsa

THE AUTHOR

Veikko V. Sorsa is Professor in Genetics at the University of Helsinki. He started with chromosome studies on plants, particularly, cytotaxonomy of fern genus *Dryopteris,* but turned in the 1960s more and more to the electron microscopy of polytene chromosomes. Besides the gene localization and ultrastructural studies, he has carried out with his colleagues a long-term project for electron microscopic revision of the reference maps of the salivary gland chromosomes of *Drosophila melanogaster*.

He received his Master's degree in biological sciences from the University of Helsinki 1954 and presented his doctor's thesis in genetics 1956. After acting as a teacher and lecturer from 1955 through 1963 at the University and Teacher Academy of Helsinki, he was a visiting Research Associate in Genetics at the University of California, Berkeley from 1963 to 1964 and also participated as a fern cytologist in the Rain Forest Irradiation Project of U.S. AEC in Puerto Rico in 1964 and 1965. From 1964 through 1976 he was first the junior and then the senior Research Associate of the National Research Council of Finland. During that period he was a visiting Research Fellow at the Max-Planck-Institut für Biologie, Abt. Beermann, in Tübingen, F.R.G. in 1968, and Research Associate at the University of California, Davis from 1974 to 1975. In 1977 through 1984 he was Associate Professor, and since 1984 a personal Professor in Genetics at the University of Helsinki. He is a Member of the Finnish Academy of Science, and of the Finnish and Scandinavian Societies of Genetics.

To Marja and Pekko

To the Memory of My Parents

TABLE OF CONTENTS

Volume I

Volume II

Chapter 15

CHROMOSOME MAPS OF *DROSOPHILA MELANOGASTER*

I. LIGHT MICROSCOPIC MAPS OF *DROSOPHILA MELANOGASTER*

A. Camera Lucida Maps of Painter

T. S. Painter has been mentioned as one of the first cytologists, who already in the early 1920s counted the diploid chromosome number 46 for the human cells.[1] Later on he turned more and more to the cytological study on *Drosophila*, publishing cytological maps of mitotic X chromosome,[2,3] and finally he became a pioneer and well known authority in cytology of giant polytene chromosomes (see Chapter 2 in Volume I).

Painter[4] was the first to realize that the constant morphology with a characteristic banding pattern in the salivary gland chromosomes could be used for chromosomal localization of the breakpoints of inversions, translocations, and other chromosomal rearrangements. He also gave the principles of how the gene loci could be mapped in the polytene chromosomes by studying these chromosomal rearrangements. The exact pairing of homologs readily shows the deleted, inverted, or translocated parts in the giant chromosomes.

Already in the first schematic map of X chromosome,[4] Painter demonstrated approximate chromosomal location of about 20 genetic characters. In another paper submitted to *Genetics* in May 1933, but published about a year later,[5] Painter presents camera lucida drawings of several inversions, particularly for demonstrating the intimate association of homologs. The absence of an inert region from the proximal part of a salivary gland X chromosome became evident by the localization of the dominant marker Bar, which in the mitotic X chromosome seems to be located in about the middle of arm. Although the localization of genetic markers was still very inaccurate in the early maps, the order and distances of genes in the salivary gland chromosome map seemed to be "in very close agreement with the crossover map concept".

Painter's second paper in *Genetics*,[6] also submitted before the first report in *Science*,[4] described the procedure used by making preparations from the salivary gland chromosomes. This paper includes a revised map of X chromosome (Figure 1)* and presents a large collection of camera lucida drawings of chromosomal aberrations, many of them with detailed descriptions. Obviously the equal pairing of homologs in the salivary gland chromosomes of both males and females lead Painter to also discuss the problems of somatic and meiotic pairing. The equal pairing of homologs in the salivary gland cells did not seem to be in accordance with the ideas of the total absence of meiotic pairing and recombination from the males of *Drosophila melanogaster*. Another problem considered by Painter was the elimination of the inert region of chromatin from the X chromosome in the salivary gland cells.

Painter's article in the *Journal of Heredity*[7] reviewed the earlier work and more contemporary attempts on polytene chromosomes, also mentioning the mapping project started by Bridges at the end of 1933. Painter summarized the results obtained at the University of Texas at Austin, including the first complete maps of the salivary gland chromosomes of *D. melanogaster* (Figure 2). The main problem, the chromosomal location of genes, is discussed at length. The importance of short deletions in exact localization of genes is pointed out, and it was shown that irradiation could be successfully used for producing small deletions.

By using the deletion mapping, for instance, the *vermilion* locus was restricted to "the

* All figures appear after the text.

left part of the broad compound band'' later known as doublet 10A1-2 according to the reference maps of Bridges. This may be considered as the first evidence ever obtained for the band location of a gene.

In a larger article submitted to *Genetics* in the summer 1934[6] Painter describes the morphology of the salivary gland chromosome 3 in *D. melanogaster* including the maps of both arms and drawings of about 50 aberrations in them. Consistently with the observations of Heitz[9] Painter describes that the chromocenter in the salivary gland cells is composed of heterochromatic proximal regions of all arms. In the mitotic interphase of somatic cells the chromocenter contains almost the whole Y and a large proportion of proximal X chromosomes. Comparison of the genetic and cytological map of the salivary gland chromosome 3 still supported a similar order of genes although the variation in distances between the loci seems greater. Comparison with cytological maps of mitotic chromosome 3 shows that proximal parts of arms around the spindle fiber attachment site apparently are involved in formation of chromocenter. The unequal distribution of irradiation breaks in chromosomes, also suspected earlier, seemed to be clearly demonstrable in the giant salivary gland chromosomes.

B. Reference Maps Compiled by C. B. Bridges

At first as a student of genetics in Morgan's school, and since 1915 as a more independent research associate of the Drosophila-group at Columbia University supported by the Carnegie Institution of Washington, C. B. Bridges participated for more than 20 years in the construction of genetic maps of *D. melanogaster.*[10] When the Drosophila-group finished the mapping project at New York about 15,000 generations of flies were handled from the start with the first strain of *D. melanogaster* collected by Payne.[11] Bridges moved with Sturtevant, Schultz, Tyler, and Dobzhansky to California and continued his studies in the Carnegie Institution of Washington resident at the California Institute of Technology, Pasadena, where Morgan had been invited to organize the Division of Biology (Volume I, Chapter 2, Section III).

In order to make use of the new type of chromosome maps introduced by Painter,[4] Bridges also started the studies on salivary gland chromosomes of *D. melanogaster.* According to Morgan,[10] C. B. Bridges was ''eminently fitted'' for chromosome mapping because ''his eyesight was unusually acute but he did not depend on this alone and spent much time in a study of best methods of illumination, the best combination of lenses, and the most suitable staining and preserving technique to bring out the fainter bands''. The original drawings combined from numerous camera lucida sketches of regions showing the best possible resolution were shown in the annual poster exhibit of the Carnegie Institution of Washington at about the same time as the Painter's maps of the whole salivary gland genome were published in the *Journal of Heredity.*[7] The reference maps of Bridges with a band cataloging system were published in the next volume of the *Journal of Heredity* in 1935[12] (Figure 3).

Most impressive proof for the possibilities of the new type of mapping was shown by the direct comparison of chromosome 4 of *D. melanogaster* in the camera lucida drawings made on the same scale from mitotic and salivary gland cells.[12] Besides the prominent accuracy, with more than 3280 lines (counting the doublets as two separate lines), the maps were provided with a reference system for identification of all bands. The reference system proposed by Bridges divides the limbs of salivary gland chromosomes into 102 sections called ''divisions'' designated by numbers from 1 to 102. Each of the five main limbs (X, 2L, 2R, 3L, and 3R) contain 20 divisions and the short chromosome 4 only 2 divisions. The divisions are started with a prominent band and divided further into 6 subdivisions, each designated with capital letters from A to F. Each subdivision starts wiih a sharp band. Thus each individual band of salivary gland chromosomes can be identified by giving the division number, subdivision, and the number of band starting from the beginning of the subdivision. In the first reference maps of Bridges[12] the numbering of bands was not yet presented because it soon became evident that many new bands could be found in the future.

Bridges[12] presents the following minimum numbers of bands for the salivary gland chromosomes of *D. melanogaster:* 537 bands for the X chromosome, 1032 bands for the second chromosome, 1047 bands for the third chromosome, and 34 bands for the fourth chromosome, which gives 2650 bands for the whole genome. In this count the doublets are listed as single bands. In later interpretations presented by C. B. and P. N. Bridges in the papers introducing the revised maps[13-17] the doublets have been taken in account giving 725 bands for the X chromosome,[13] 660 bands for 2R,[14] 542 bands for 3L,[15] 725 bands for 3R,[16] 584 bands for 2L,[17] and about 50 bands for the chromosome 4.[17]

These interpretations give a total number of about 3286 bands for the whole genome. (The table included in the report of revised 2L[17] apparently gives an erroneous number of bands for 3R.) For the reference maps[12] C. B. Bridges used moderately stretched salivary gland chromosomes having an average total length of about 1180 μm (X = 220 μm, 2L = 215 μm, 2R = 245 μm, 3L = 210 μm, 3R = 275 μm, and 4 = 15 μm). According to Bridges[11] this length represents about 150 times the total length of gonial mitotic chromosomes (about 7.5 μm).

Independently from the "polyteny hypothesis" proposed by Koltzoff[18] C. B. Bridges also came to a conclusion that "the large size of the salivary gland chromosomes is partly due to their being compound structures. Each of the fused maternal and paternal homologues consists of eight chromonemata or gene-strings derived from the corresponding chromosome of the gamete by successive divisions without complete separation of the division products". Bridges also noticed that the free ends of salivary gland chromosome arms presented "characteristically narrowed terminal regions" suggesting that the narrowed ends of limbs represent a lag of one division in the gene-string. Thus Bridges' view of polytenization was in good accordance with recent models of the branched structure at the chromosome ends.[19]

Bridges[12] described several direct and reversed repetitions in the banding pattern showing similar or identical morphology. As clearest examples of symmetrical reversed repeats in the band sequence he mentioned the "turn-back" in division 36 and the "shield" in 30A. In the "shield" he pointed out the ectopic fusion of the material of bordering heavy bands along the surface of the chromosome. Bridges also proposes that the four-banded capsule-like groups, e.g. in 25A and 56F, as well as the reversed symmetries found around certain weak spots, are reversed repeats. As an example of a more complicated combination of direct and reversed repeats he mentioned the region showing ectopic pairing in the area 32E-35A of 2L. Expectedly the older and longer repeats are structurally more changed during the evolution than the younger and shorter repeats. As examples of very short repeats of even one single band Bridges mentioned certain capsule-like band complexes.

Bridges also revives his old hypotheses, which he had presented already almost 20 years earlier, that duplications at different levels of chromosome structure are evolutionarily important. According to his concept, only duplications offer "a method for evolutionary increase in lengths of chromosomes with identical genes which could be subsequently mutate separately and diversify their effects". Bridges emphasized that "the demonstration that certain sections of normal chromosomes have actually been built up in blocks through such repeats goes far toward explaining species initiation". Later evidence derived from different sources has strongly supported the role of tandem duplications of genetic material at different levels as an important basis of evolution.[20-23]

In order to easier utilize the reference maps for recognition of chromosome arms Bridges gives a number of structural "landmarks" which he suggested are useful to be learned. Bridges describes following easily recognizable sites, starting from the X chromosome: "Puff" at the subdivision 2B, the "four brothers" at 9A, the "weak spot" at 11A, the "chains" in 15B and D, and the "turnip" in 16 (A-C). In the second chromosome he also mentioned several typical recognition sites: The ''dog collar'' at 21C-D, the ''shoe-buckle'' at 25A, the ''shield'' in 30A, the ''goose-neck'' at 31B-F, the ''spiral loop'' in region 32-35, a ''turn-back'' in 36 and the ''basal loop'' in region 37-39 all in the left limb of chromosome 2. The right arm of the salivary gland chromosome 2 can be recognized from the ''onion''-like base and the ''huckleberry'' tip. From the left arm of the third chromosome Bridges describes four ''landmarks''; the ''barrel'' at region 61C-F, a ''ballet skirt'' at 68BC, ''chinese lanterns'' at 74-75, and ''graded capsules'' in 79C-E. The right arm of chromosome 3 is the longest one and has a large ''cucumber'' at the region 81-83D, a ''duck's head'' in the region 89E-91A, and a ''goblet'' tip. In the explanations of his photomaps of the salivary gland chromosomes of *Drosophila melanogaster* Lefevre[24] has still added some descriptive terms; the ''goggles'' for the bands in area 50A-C, a ''barrel'' at division 47, and ''road apple'' at region 85F.

C. The Maps Compiled by King

The revised versions of the reference maps of the salivary gland chromosomes of *D. melanogaster* published by King[25,26] were compiled and redrawn on the basis of the maps of Bridges[12-17] and Slizynski.[27] The cytological maps were also provided with the genetics maps compiled according to Lindsley and Grell.[28] In the latest version of King's maps[29] the previous map of the fourth chromosome has been replaced with a new one compiled by B. Hochman (Figure 4).

D. The Photomaps of Lefevre

Already more than a half of a century the reference map drawn by C. B. Bridges (1935)[12] has served as basic standard for the cytogeneticists working on *D. melanogaster*. Invention of phase-contrast optics and development of automatic photomicrographing equipment have made it possible to construct photographic maps of the salivary gland chromosomes. If compared with the camera lucida method the photomicrographing certainly eliminates the more or less subjective standpoints of the investigator. On the other hand the focus plane chosen by the photographer from several alternative focuses may give a view differing from the combined multifocal picture obtained by camera lucida drawings. Thus, although being more objective, a photograph may miss some detail which can be compiled only by looking at the giant polytene chromosome at numerous focal planes. Current knowledge about the structural organization of chromosomes certainly helps us to save the most essential components of chromatin and to prevent obvious artifactual changes caused by the preparation procedure.

The modern techniques for making preparations of polytene chromosomes have been described by Lefevre[21] in his article introducing the photographic representation of the salivary gland chromosomes of *D. melanogaster*. The photomaps, although very similar to the maps of Bridges[12] are also resembling more like normal preparations seen in the light microscope, which greatly helps the use of the reference maps. Besides the photographic representation, the salivary gland chromosomes are well-characterized and compared with the camera lucida map of Bridges (1935).

According to the interpretation of Lefevre[21] the total number of bands in the first reference map of C. B. Bridges (1935)[12] is 3540 of which 725 bands are in the X chromosome, 1320 bands in the chromosome 2, 1450 bands in the chromosome 3, and 45 bands in the chromosome 4. Thus there seems to be several different ways to make band counts from the same maps. The lowest number of bands (2650) was given by C. B. Bridges.[12] From another count summarized from the papers of C. B. and P. N. Bridges[13-17] introducing the revised maps we can get 3286 bands for the whole genome. The more recent analysis of Lefevre,[21] as it was mentioned above, has revealed the highest number; 3540 mapped bands for all salivary gland chromosomes of *D. melanogaster*.

Even the best possible photographs are not able to show more than about 50 to 70% of the bands drawn by C. B. Bridges in his first map.[12] Thus the light microscopic analyses have not been able to verify all the bands depicted by Bridges even in the first reference map. Lefevre[21] has particularly criticized the tendency of Bridges to see and describe too regularly many of the direct and reversed repetitions he proposed to be present at different levels in the salivary gland chromosomes. Lefevre has demonstrated various aspects, e.g. of the effects of stretching of salivary gland chromosomes, by means of photomicrographs. It is known that the stretching may increase the number of recognizable faint bands. According to Bridges it also uncovers double-like organization in many bands. As shown by the photomicrographs of Lefevre especially the faint bands seem to become broader i.e. stretched, and more separated. That is probably the reason why they are better observable in the light microscope (see Figure 3). Unfortunately, due to the low resolution of even the best micrographs, for instance, the real effects of stretching on the faint bands and interbands cannot be demonstrated.

By combining together a great number of carefully selected photomicrographs showing the salivary gland chromosomes equally stretched as in the camera lucida map of Bridges,[12] Lefevre has succeeded to create an excellent guide map to be used together with the reference map.[21] Besides the "landmarks" described by C. B. Bridges[12] Lefevre has added in the explanations of maps a number of "landmark constrictions" and "landmark puffs" to help the users of the maps (see Figure 3).

II. REVISED REFERENCE MAPS

During the years 1934 to 1937 there was a "chromosome revolution" in *Drosophila* research, if to use the term adapted from C. D. Darlington's historical review in the Sixth International Chromosome Conference.[30] Bridges[31] describes that period with the following expressions: "Before the importance of the salivary gland chromosomes was pointed out by Painter, more could be learned about the distribution and behavior of chromosomes and about their internal differentiation by counting the flies than by looking at the germ cells themselves". But it should not be forgotten, as pointed out by Bridges[31] and Morgan,[10] that it was the already existing genetic map which gave an instant meaning to the banding pattern found in the salivary gland chromosomes of *D. melanogaster*.

After getting ready the reference map[12] C. B. Bridges still spent much time by improving the preparation methods of polytene chromosomes in order to find an optimal procedure capable of giving maximal number of bands, particularly, of the faintest bands. Soon it became evident that how useful the first reference maps had been, they were still very inaccurate for exact studies of small chromosomal aberrations, and that the revision of reference maps was necessary.[31]

Bridges published an accurate camera lucida drawing showing the distal divisions of 2R chromosome of *D. melanogaster* with a detailed genetic map also revised according to the latest data[31] (Figure 5). The map of the distal regions of 2R was evidently the first high-resolution analysis of any chromosome region. Bridges describes the cross bands as "solid, inelastic and disc-like, while the material between bands is very elastic — drawing out to at least three times its lax length without apparent disruptive change in its structure". From the results of Feulgen-staining of salivary gland chromosomes Bridges stated that the "bands are highly charged with nucleic acid while the interband regions are relatively free of nucleic acid". According to the current knowledge of the structural organization of chromatin the length of DNA released from the nucleosomes is roughly six to seven times the axial length of nucleosomes in the chromatin fiber (see Volume I Chapter 8).

By the revision of maps Bridges was able to find very high numbers of doublets that he suggested "their nearly universal appearance".[31] Particularly, the heavy bands were frequently seen as double-structures, but according to Bridges, "even the medium and faint lines seem to be clearly doublets in the great majority of cases". The revised map of the distal 2R also introduces two suggestive symmetrical reversed repeats, one in region 56F-57B, and another in region 58 E-F with a swollen and light-staining puff area. Electron microscopic analysis of 56-57 have shown that the halves of assumed reversed repetition are not quite identical in the ultrastructural level.[32] This does not exclude their repeat origin, but suggests that some structural changes have also occurred in the region.

A. Revised Map of X Chromosome

Bridges started the revision of the reference maps of entire arms with publishing a new map of the X chromosome.[13] The revision of maps was based on camera lucida drawings from carefully chosen permanent preparations having much better transparency than the temporary preparations used previously. Bridges describes the method he developed for making the permanent slides. The essentials of the improved preparation techniques used by Bridges were the following: limited number of larvae per culture were raised to maximum size at low temperature (17 to 19°C) by adding extra yeast. The salivary glands of prechilled larvae were prepared in chilled Ringer, and the glands were fixed with the aceto-carmine stain, squashed on albumen-coated slides, dehydrated in vapor of 95% alcohol and mounted in euparal from a 95% alcohol bath after removing the cover slip used for squashing.

Instead of replacing the old map of X chromosome with the new one, Bridges suggests that the previous map (1935) could be further used as a guide map, but all the analyses of small aberrations requiring ultimate accuracy should be carried out by using the revised version.[13] According to Bridges[13] the revised map shows 1024 bands for the X chromosome. However, as pointed out by Lefevre[24] the total number of lines depicted in the revised map is only 1012 if summarized from the band numbers given on the map per each subdivision. Thus the total number of bands for the X chromosome was increased from 725 to 1012 by the revision. The main reason for this about 40% increase in the number of bands is apparently in greater number of doublets. In the new map of X chromosome Bridges has depicted more than 50% of bands as doublets. There are some divisions, like 4, 14, 17, and 18 in which from 80% to more than 88% of bands are drawn as doublets. Evidently the more extensive stretching of chromosomes made it possible to detect more doublets and faint bands. The length of the X chromosome in the first map of Bridges was only about 220 μm but 414 μm in the revised map. Four subdivision borders in the map of X chromosome between 2C and D, 7C and D, 14A and B, and 16A and B were changed by the revision of reference map.

In the revised map of X the chromosomal location of 18 genes of the genetic map has been pointed in the cytological map. Some of them, for instance, the *white*-locus is already localized very exactly, as shown by the later studies on the chromosomal site of this gene.[33-35] Bridges also explains the polytene organization of salivary gland chromosomes: "The salivary gland chromosome is thus essentially a bundle of parallel chromonemal threads, similar to that of ordinary chromosomes". But the total number of parallel strands he estimated to be much too low, if compared with the more current evidence of the degree of polyteny.

Bridges localized the attachment site of "the huge bag like nucleolus" in the region 20 B-C at the proximal end of X, and listed a number of "difficult regions" e.g. in divisions 11, 12, 19, and 20. He suggested also possible existence of some reversed repeats in X, e.g. in the regions 2B, 3C, 8B, 9A, and 11A.

B. Revision of the Map of 2R Chromosome

C. B. Bridges also started the revision of the other chromosomes with new camera lucida drawings of the entire chromosome 2R.[10] The map was published together with his son

P. N. Bridges, who prepared the paper for press after the early death of C. B. Bridges in 1938[10,14] (see Volume I, Chapter 3, Section III).

The length of the 2R chromosome in the new map is depicted to be 446 μm showing 1136 numbered cross bands. The corresponding values in the first reference map were 245 μm and 660 bands for the salivary gland 2R chromosome.[12] The increase of total length was explained to be due to choice of larger chromosomes and more stretched regions used for the mapping. The authors remind that the faint bands or doublets missing from the preparations when making comparisons with the revised maps are most probably lost because of the insufficient stretching of chromosomes.

Besides the most proximal, chromocentric division 41, also some other regions of the chromosome arm 2R are mentioned to be difficult. According to the authors such regions are e.g. certain puffed areas like 42A, 43E, and 47C, but also certain heavy band groups like 52C, 55C, and 58E. Also the regions composed of mainly faint or less stainable bands like in the divisions 43, 46, 52, 54, 56, and 57 could be difficult to interpret in preparations. The authors mention a possible repeat in area 56F-57B[14] accordingly to the earlier report of C. B. Bridges.[31] Occasionally this region seems to show ectopic pairing affinity between the heavy bands.[32]

C. Revised Maps of Chromosome 3

The third supplement to the reference map of *D. melanogaster* of C. B. Bridges[12] was the revised map of the salivary gland 3L chromosome published by P. N. Bridges 1941.[15] The revision also includes some changes in section borders. To start the sections with more distinctive bands one division border (78/79) and 17 borderlines of subdivisions have been moved by the revision of the map of 3L. Three new subdivision lines were added in the difficult proximal division 80 by the revision. The new borderlines were depicted with dashed lines in all of the revised maps.

The length of chromosome 3L in the revised map is 424 μm, which is more than twice the length (210 μm) of the same chromosome in the previous reference map.[12] Respectively, the number of bands was increased about 63% from 542 to 884 in the revised map of 3L. Besides the chromocentric region of this chromosome arm, Bridges has particularly named certain other easily recognizable diffuse regions along this arm. Those are e.g. the faintly staining regions in 63AB, 67AB, 70A-C, 74, 74AB, 76CD, and 79EF. Other recognizable landmarks added by P. N. Bridges are e.g. the wide or puffed areas in 64C, 68C, 71C-E, 72D, 74D, and 79EF.

The revised map for the right limb of the third chromosome was also published in 1941 by P. N. Bridges[16] as the fourth supplement to the reference maps of the salivary gland chromosomes of *D. melanogaster*.[12] By using the methods also described in the earlier papers of revised maps, P. N. Bridges completed the revision of the third chromosome. Excluding some minor corrections no changes of division or subdivision borders were needed in the revision of 3R. For some reason C. B. Bridges,[12] when creating the reference system, reserved two whole divisions 80 and 81 for characterizing the banding pattern in the chromocenter region of the third chromosome. By the revision of the map of this chromosome P. N. Bridges succeeded to map the banding pattern in the proximal subdivisions 80D-F of 3L but not more than the last subdivision F of division 81 at the proximal end of 3R.[15,16] Thus the first complete division in the chromosome arm 3R is 82.

The length of 3R, the longest limb of genome is 275 μm in the old (1935) map showing 725 bands.[12] In the new, revised map P. N. Bridges has depicted the chromosome arm 3R as 519 μm long and containing 1178 bands.[16] Thus the new map shows about a 62% increase in the band numbers, and about an 89% increase in total length, if compared with the reference map of C. B. Bridges.[12] Besides the typical landmarks of chromosome 3R given already in the explanations of the first reference map P. N. Bridges added some other, like the puff in 85F and the ''road apple'' puff in the terminology of Lindsley, (see Lefevre[24]). P. N. Bridges also mentioned as typical ''landmark'' regions ''the series of heavy bands in 93E and F'', the ''bulge'' at 94D, the general ''slim tapering of section 95'' and the heavy doublets in 100B, just before the ''goblet'' tip of the chromosome.

D. Revision of the Map of Chromosome 2L

The fifth revision of the reference maps of the salivary gland chromosomes of *D. melanogaster* published by P. N. Bridges in 1942[17] concerns the left limb of the second chromosome. The revision of this chromosome arm was left to last because of its complicated structure, at least in certain regions. The length of 2L was increased by revision about 72% from 215 μm in the old map, to 370 μm in the new map of this limb.[17] Correspondingly, the total number of bands depicted in the maps was increased about 38% being according to Bridges 584 in the old map while 804 bands can be counted from the revised map of 3L.

Apart from the landmarks given already in the explanations of the old reference maps,[12] some new recognizable and constant sites were proposed by P. N. Bridges to help the users of the maps; the ''salpas'' at 32, the dilation in the area of subdivisions 35AB, and groups of heavy bands in certain regions like 34E and F, 36D, and 39D. In this last revision[17] P. N. Bridges also summarizes the results of band counts and makes some comparisons with the lengths of arms in relation to their band numbers.

Bridges also discusses the problems concerning the location and number of genes in relation to single and double bands found in the salivary gland chromosomes. As a genetic evidence for tandem repeat origin of doublets, P. N. Bridges has mentioned the location of the related genes Star and asteroid. Both of these genes were localized in the doublet 21E1-2 by Lewis.[36] Star apparently locating in one and asteroid in the other half of this doublet. Structural background of the larger repeats evidenced by numerous ectopic attraction sites in the proximal 2L are also discussed in the last revision published by P. N. Bridges.

The revised reference maps of chromosomes 1-3 (C. B. and P. N. Bridges), as well as the revised map of the 4. chromosome (B. Hochman) of *D. melanogaster* are shown as division maps in Figure 1 in Chapter 16 together with the new map drawn on the basis of electron micrographs of thin-sectioned squashes of salivary gland chromosomes.

E. Revision of the Map of the Fourth Chromosome

The revision of the map of the fourth chromosome was carried out in 1944 by Slizynski.[27] Based on eight camera lucida drawings and photomicrographs of the salivary gland chromosome 4 of *D. melanogaster* Slizynski composed a revised map in the same scale previously used by C. B. and P. N. Bridges in their revised maps of other chromosomes. Slizynski presents a map also of the short left arm of the chromosome 4, the existence of which had been mentioned only occasionally.[27] In the new map the number of bands in the dominant right arm of chromosome 4 was increased from 51 bands depicted in the reference map of C. B. Bridges 1935 up to 111. A reason for the great increase in number bands in the right arm was that 40 bands were depicted as doublets in the new map. The subdivisions 101A-D of the left arm of chromosome 4 comprises 8 double bands and 8 singlets according to the revision of Slizynski. The constricted doublet 101 D3-4 apparently represents the spindle attachment region in chromosome 4.

However, the number of bands given by Slizynski in the revised map of the fourth chromosome has been considered much too high and probably erroneous.[24] More recently B. Hochman, an authority in cytogenetics of the fourth chromosome, has compiled a new and more reasonable revised interpretation of the banding in the salivary gland chromosome 4 of *D. melanogaster* (see Figure 4).

III. MAPPING OF THE POLYTENE CHROMOSOMES OF OTHER TISSUES

In accordance with the results obtained from other Diptera,[37-39] Slizynski described an essentially identical banding pattern in polytene chromosomes prepared from different tissues of *D. melanogaster.*[40] On the basis of detailed mapping of the same region from the polytene chromosomes derived from different tissues it was concluded that most of the variation found in the number and structure of bands was apparently caused by differential effects of fixation and stretching on polytene chromosomes by the preparation process. But some variation was also caused by the different type of genetic activity in certain bands appearing in the degree of decondensation and puffing of bands.[41]

An opposite view[42,43] suggesting an occurrence of more or less totally different and specific banding pattern in polytene chromosomes derived from different tissues has recently got new support. The new ideas about the possible location of active genes in linear DNA of interband regions in the polytene chromosomes[44-46] suggesting a dynamic variation of band interband composition in polytene chromosomes derived from different tissues and from different developmental stages[47] are more consistent with the latter view (see Volume I, Chapter 12).

Similarly, the variation in the replicative organization like in the length of replicons, primarily determined by the number of active replication origins, may also have an effect on the chromomeric and banding pattern. Partly for technical reasons, the banding pattern of salivary gland chromosomes seems to differ from that found in the highly polytenized chromosomes of ovarian nurse cells in *Calliphora*,[48] and of the pseudo nurse cells of ovarian tumor mutant strains in *D. melanogaster.*[49,50] Apart from the evidently differential genetic activity and puffing pattern, the chromomeric pattern of germ line cells may be different from that of highly specialized cells of larval salivary glands. The problems of phylogenetic and tissue-specific variation in banding pattern of polytene chromosomes have been recently discussed in a review by Richards.[51]

Comparison of the banding pattern of polytenized fat body chromosomes and of salivary gland chromosomes of *D. melanogaster* have shown that at least the prominent marker bands can be recognized in both of them.[40,52] However, due to differential stretching of the same regions in highly and less polytenized chromosomes, some variation appears both in the lengths of divisions and in distances of adjacent marker bands. An example of the light microscopic structure and mapping of the polytene chromosomes of pseudo nurse cells of *D. melanogaster* is shown in Figure 1, Chapter 2.

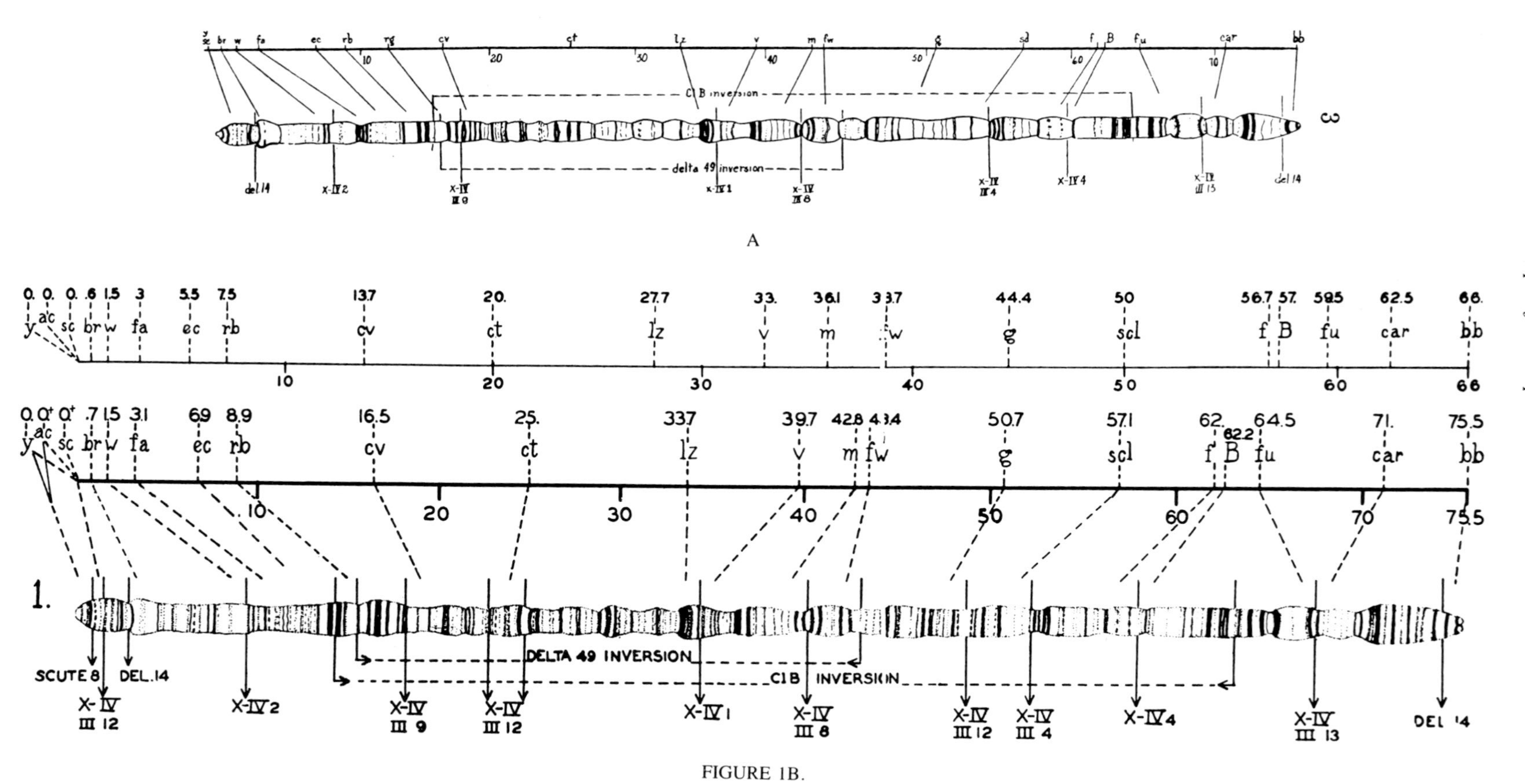

FIGURE 1B.

FIGURE 1. Painter's first map of the X chromosome of *Drosophila melanogaster* (A) already shows the ends of two inversions and a comparison with the genetic map of the X chromosome. (From Painter, T. S., *Science*, 78, 585, 1933. With permission.) Painter's revised map for the X chromosome (B) gives a more detailed comparison of the genetic and cytological maps. (From Painter, T. S., *Genetics*, 19, 448, 1934. With permission.)

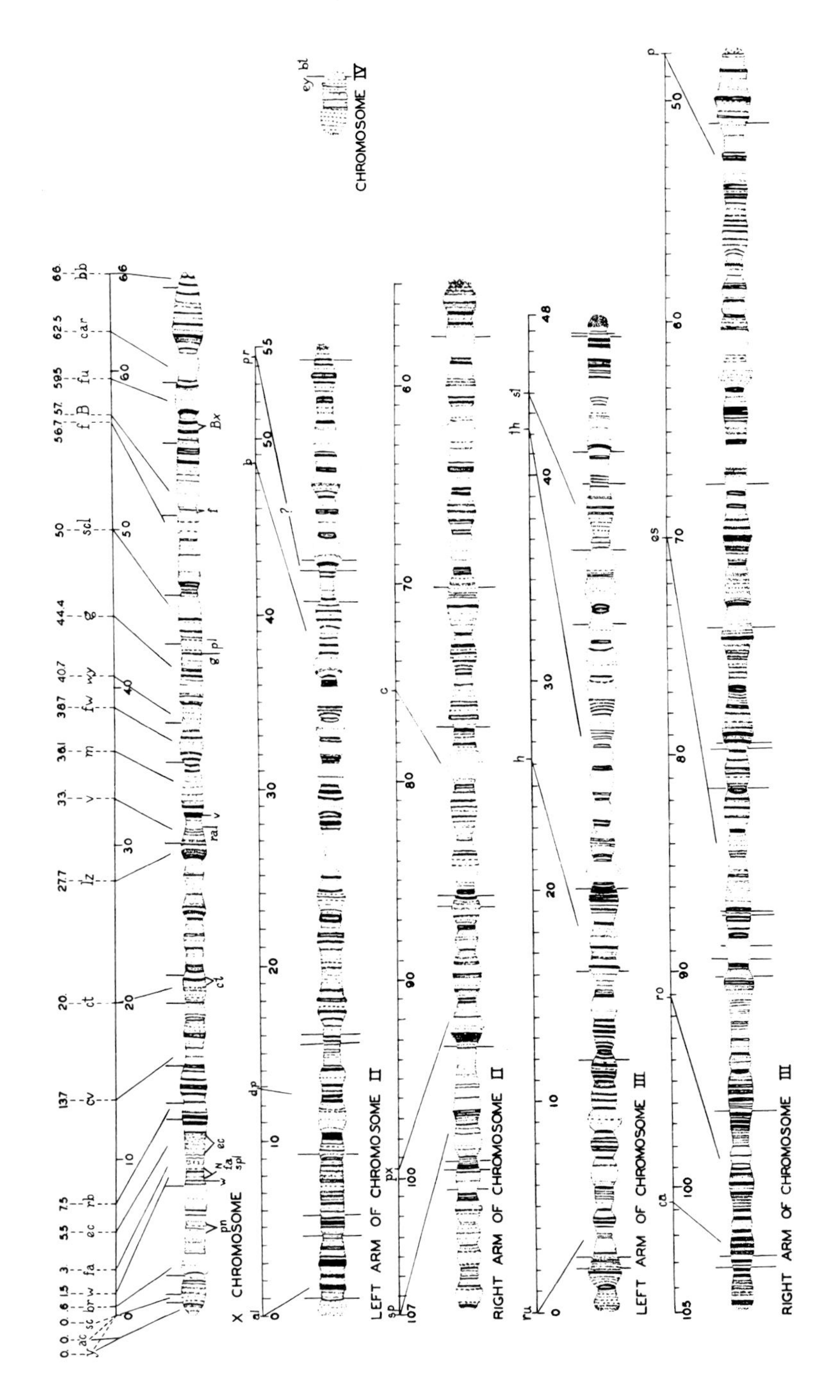

FIGURE 2. Painter's maps of the salivary gland chromosomes of *D. melanogaster* show comparisons of genetic and cytological maps of all the long chromosomes, and gives already very accurate chromosomal sites for some genes in the chromosomes X and four. (From Painter, T. S., *J. Hered.*, 25, 465, 1934. With permission.)

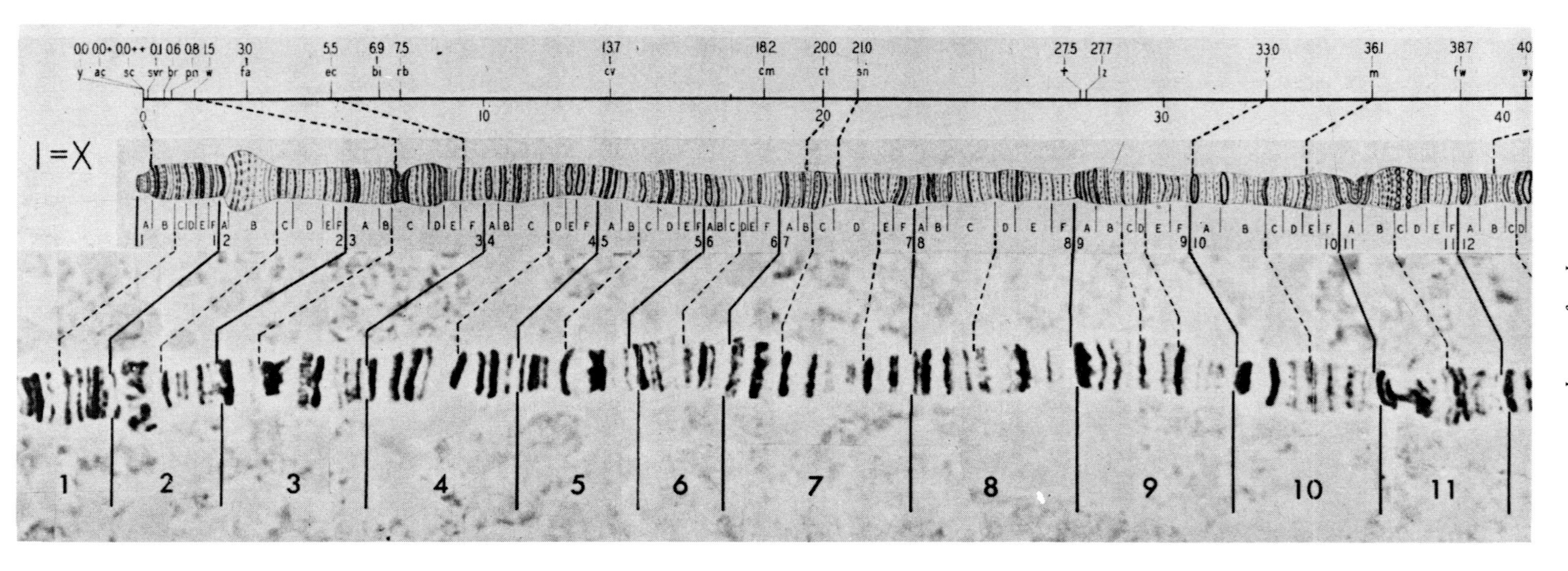

A

FIGURE 3. (A) The reference map of C. B. Bridges with a key to the banding in the salivary gland chromosomes of *D. melanogaster* (From Bridges, C. B., *J. Hered.*, 29, 11, 1938. With permission.) (B) The photographic map compiled by G. Lefevre, for interpreting the reference maps of Bridges. (From Lefevre, G., Jr., in *The Genetics and Biology of Drosophila,* Vol. 1a, Ashburner, M. and Novitski, E., Eds., Academic Press, New York, 1976, 31. With permission.)

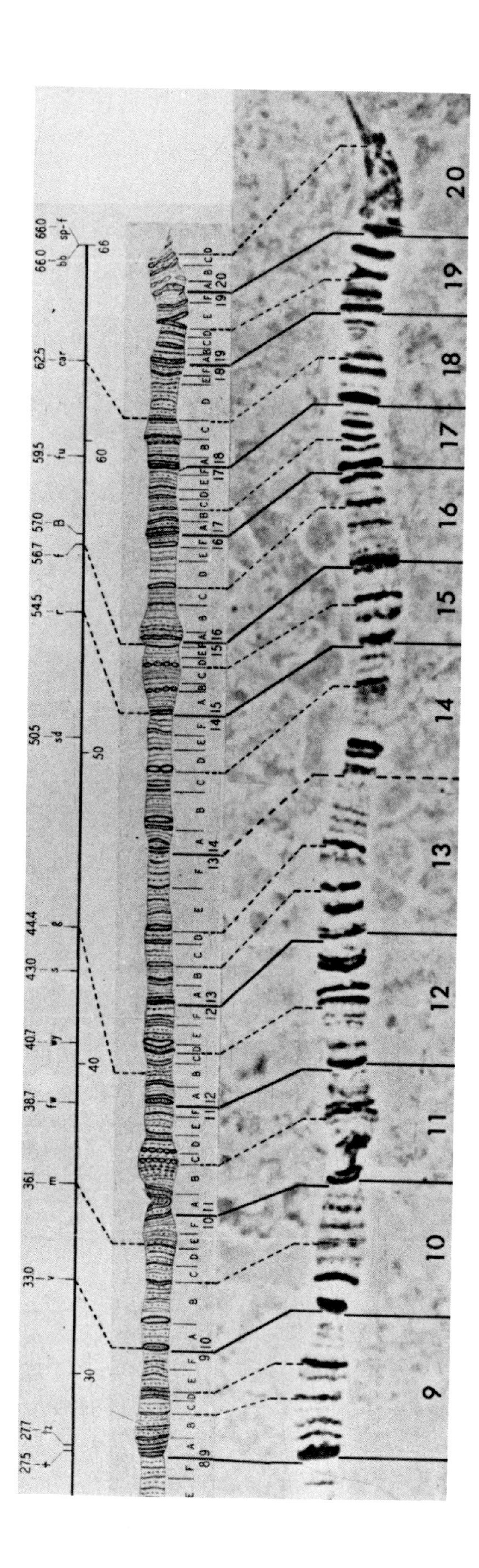

FIGURE 3A (continued)

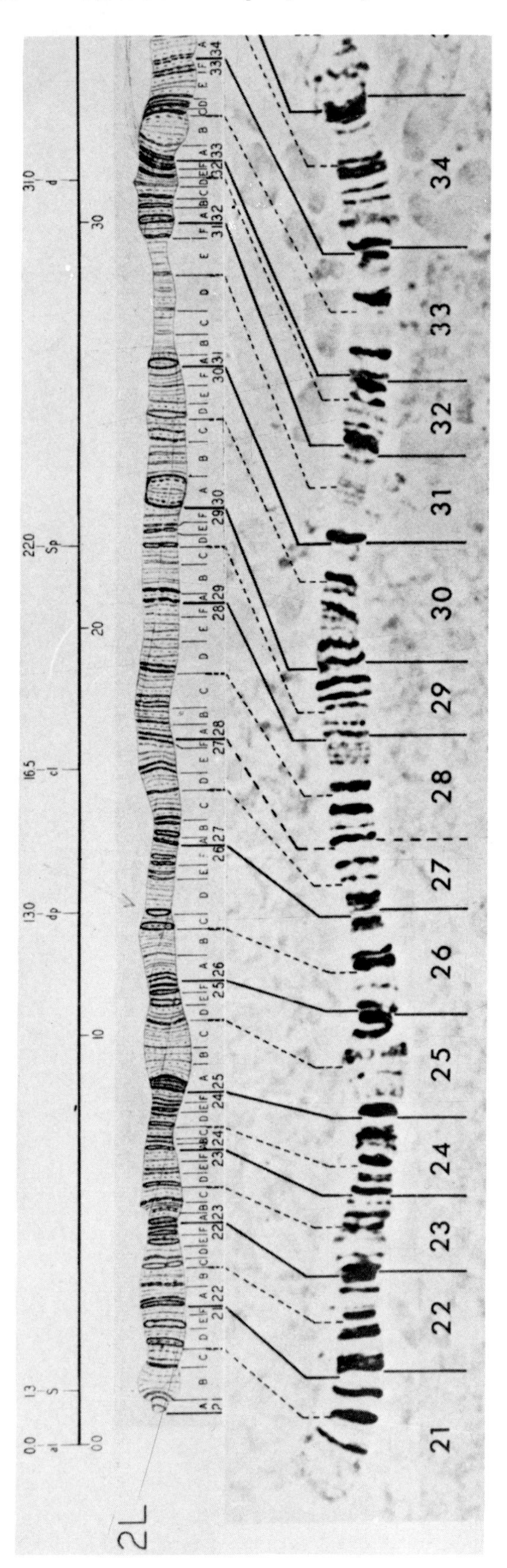
2L
0.0 al
13 S
13.0 dp
16.5 cl
22.0 Sp
31.0 d
0.0
10
20
30
21
22
23
24
25
26
27
28
29
30
31
32
33
34

FIGURE 3B

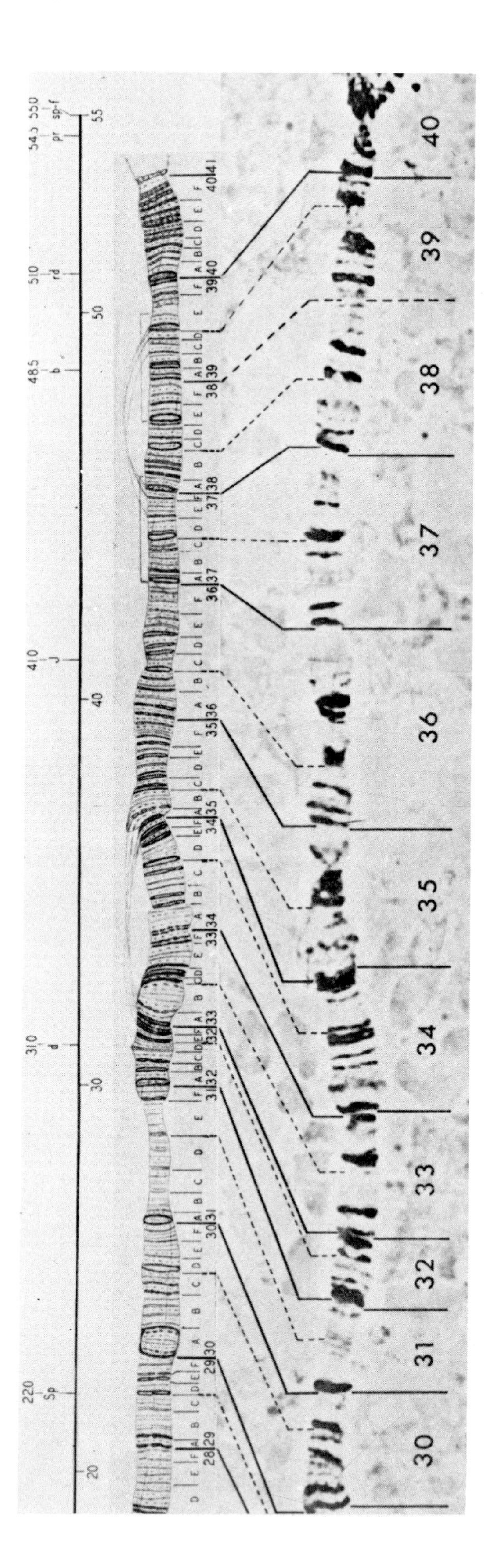

FIGURE 3B (continued)

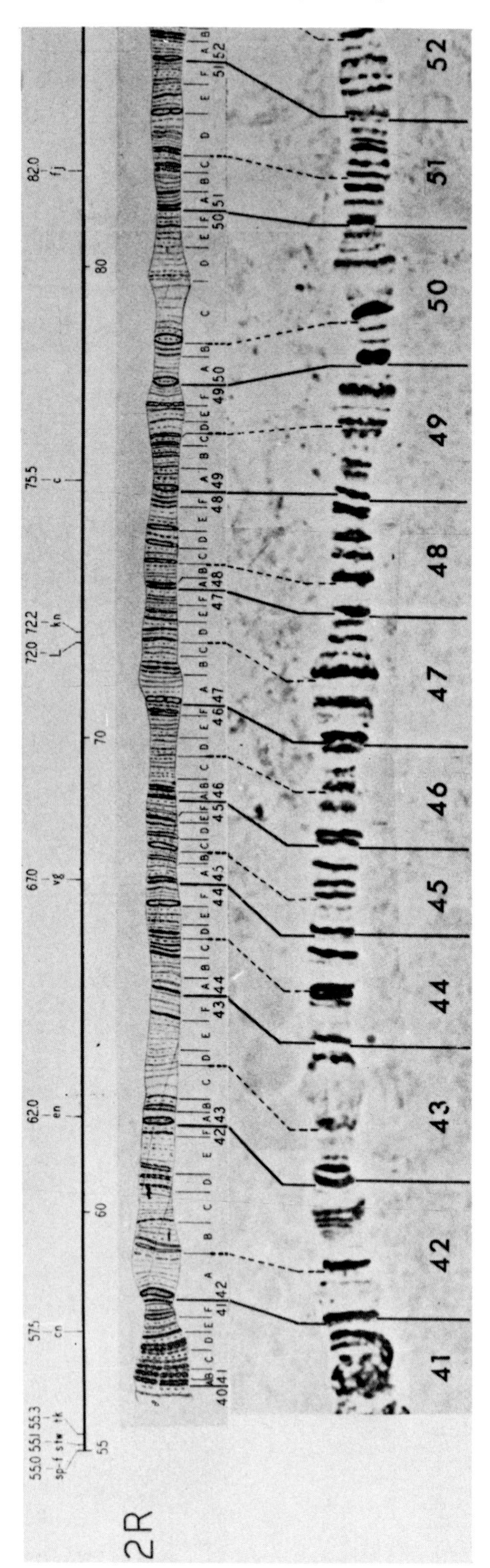
2R
55.0 55.1 55.3
sp-f stw tk
57.5
cn
62.0
en
67.0
vg
72.0 72.2
L kn
75.5
c
82.0
fj
55
60
70
80
41
42
43
44
45
46
47
48
49
50
51
52

FIGURE 3C

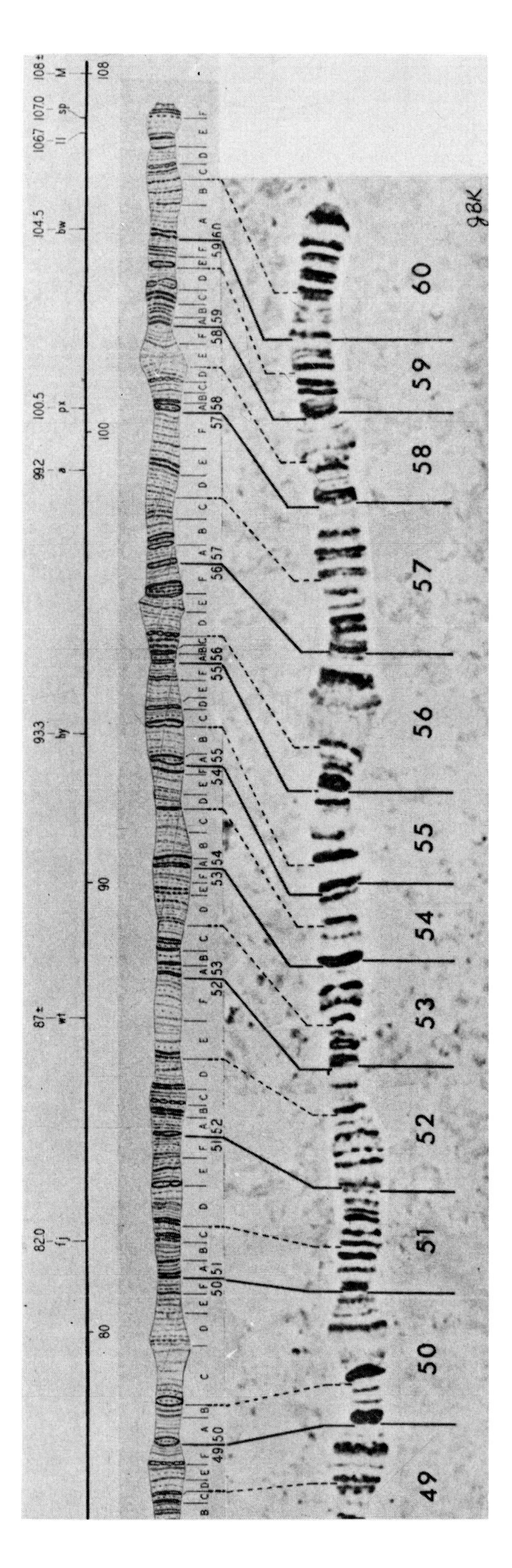

FIGURE 3C (continued)

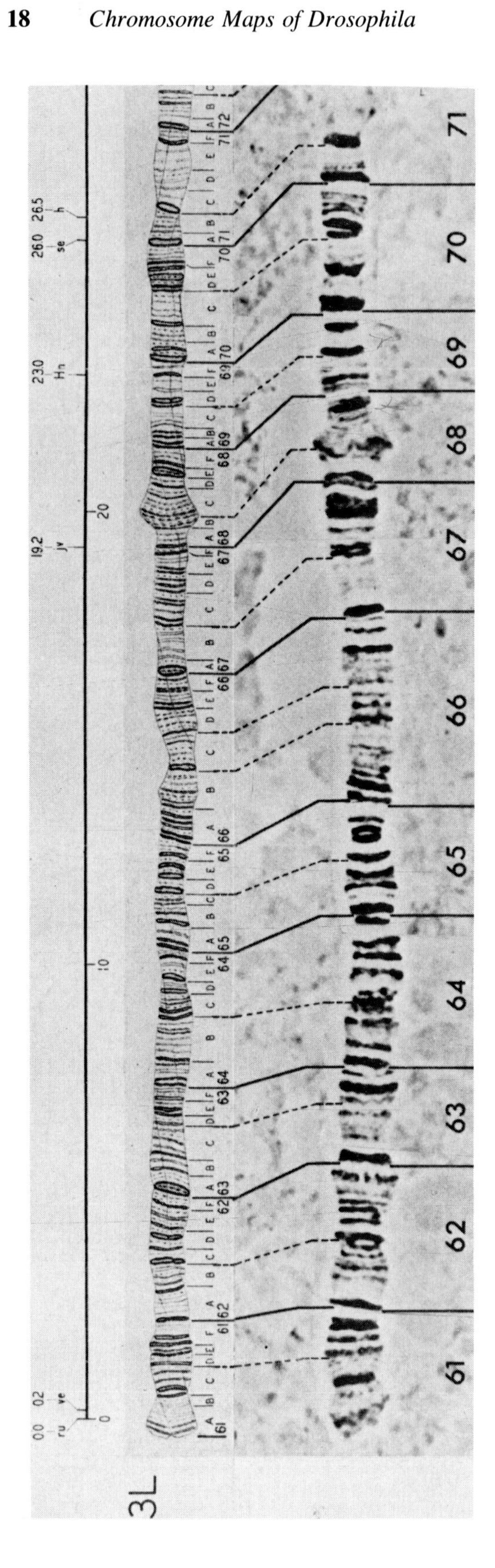

FIGURE 3D

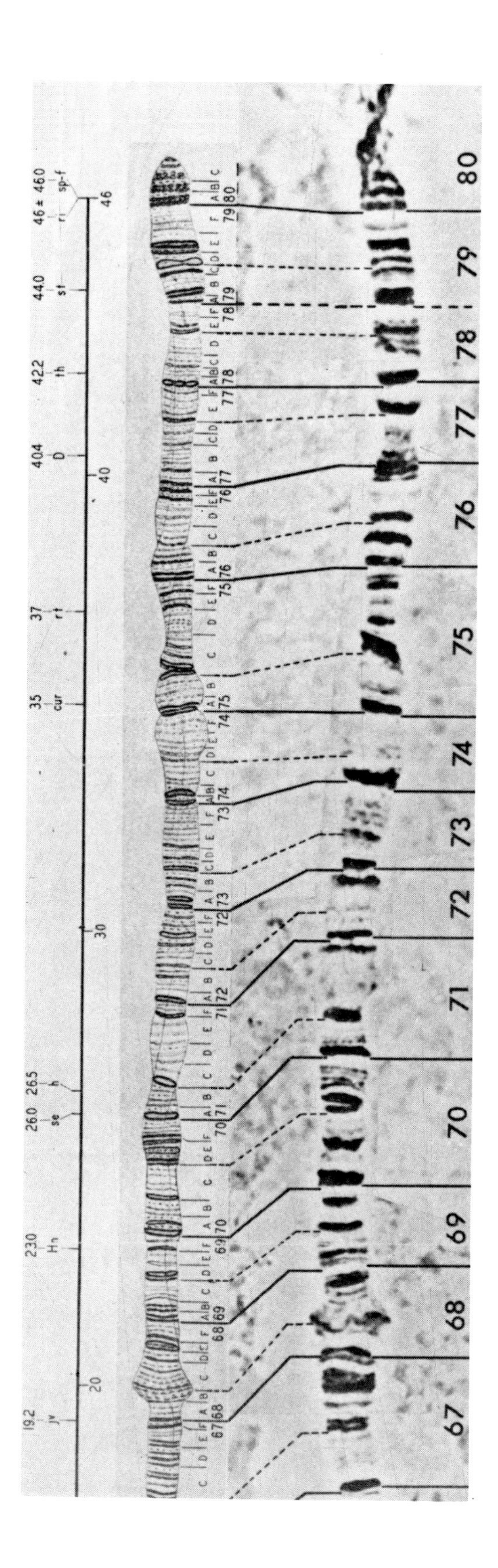

FIGURE 3D (continued)

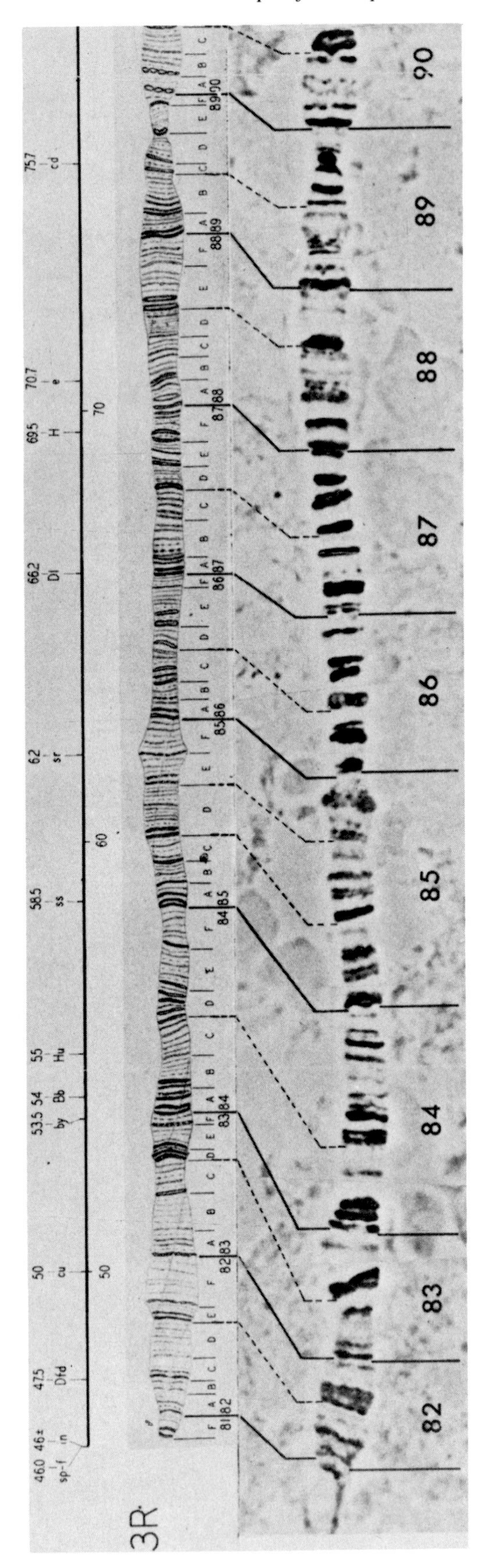
3R
46.0 46± 47.5 50 53.5 54 55 58.5 62 66.2 69.5 70.7 75.7
sp-f in Dfd cu by Bb Hu ss sr Dl H e cd
50 60 70
81|82 82|83 83|84 84|85 85|86 86|87 87|88 88|89 89|90
82 83 84 85 86 87 88 89 90

FIGURE 3E

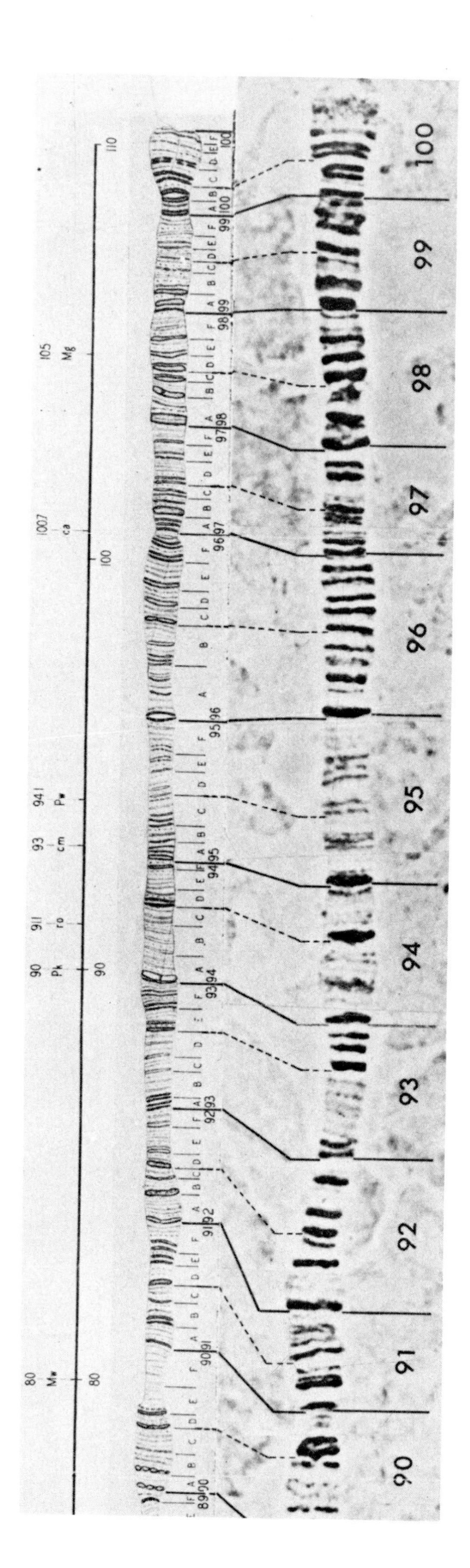

FIGURE 3E (continued)

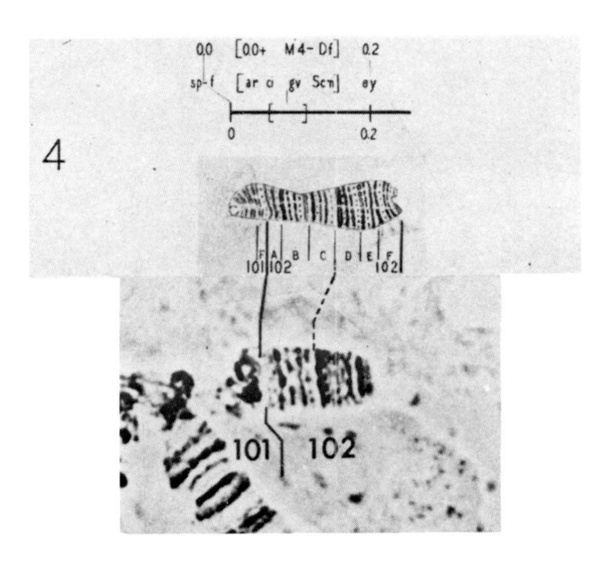
4
0.0
[0.0+ M4- Df]
0.2
sp-f
[ar ci gv Scn]
ey
0
0.2
F A B C D E F
101 102
102
101
102

FIGURE 3F

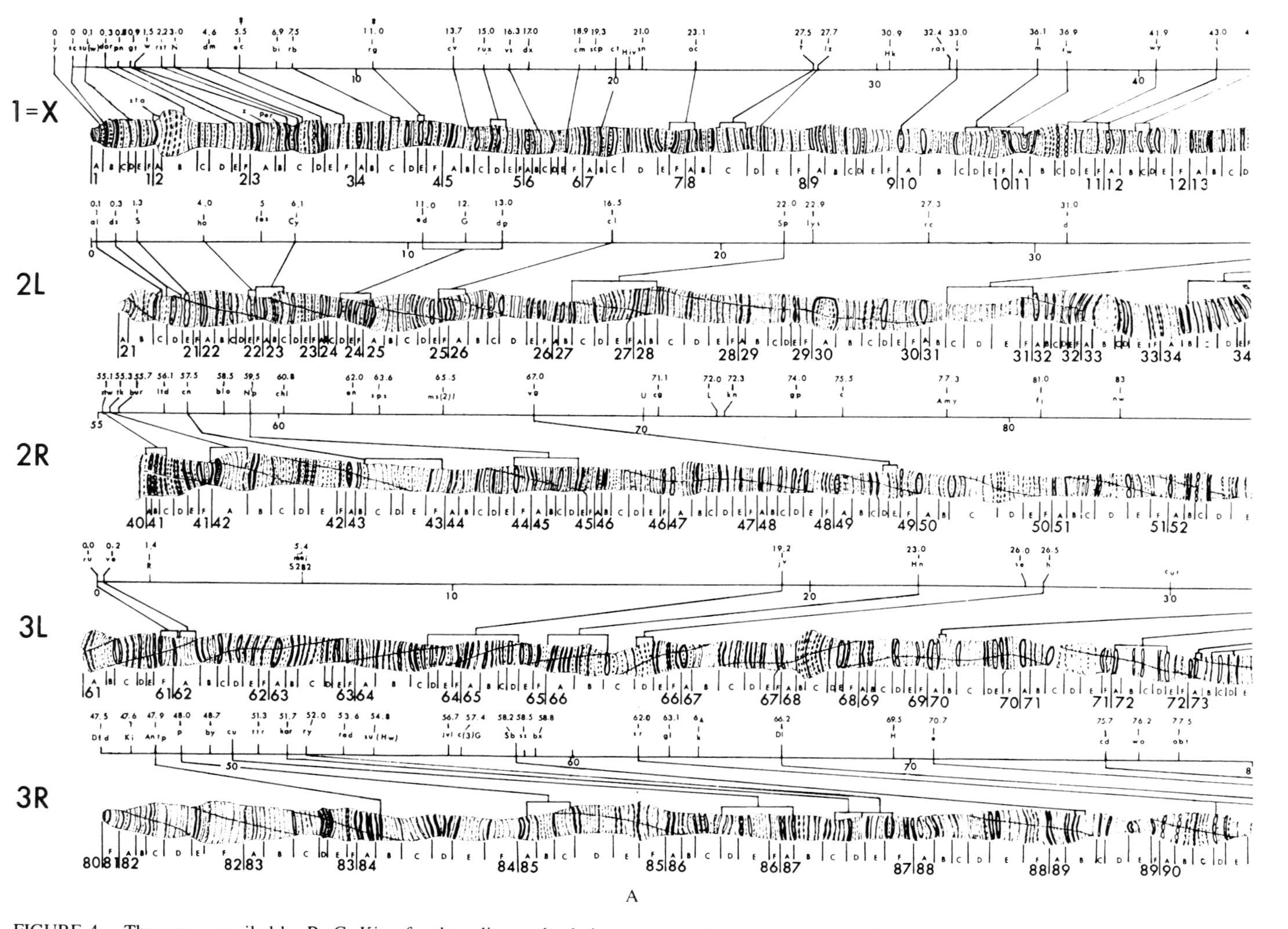

FIGURE 4. The map compiled by R. C. King for the salivary gland chromosomes of *D. melanogaster* with B. Hochman's new map for the chromosome 4. (From King, R. C., in *Handbook of Genetics*, Vol. 3, Plenum Press, New York, 1975, 625. With permission.)

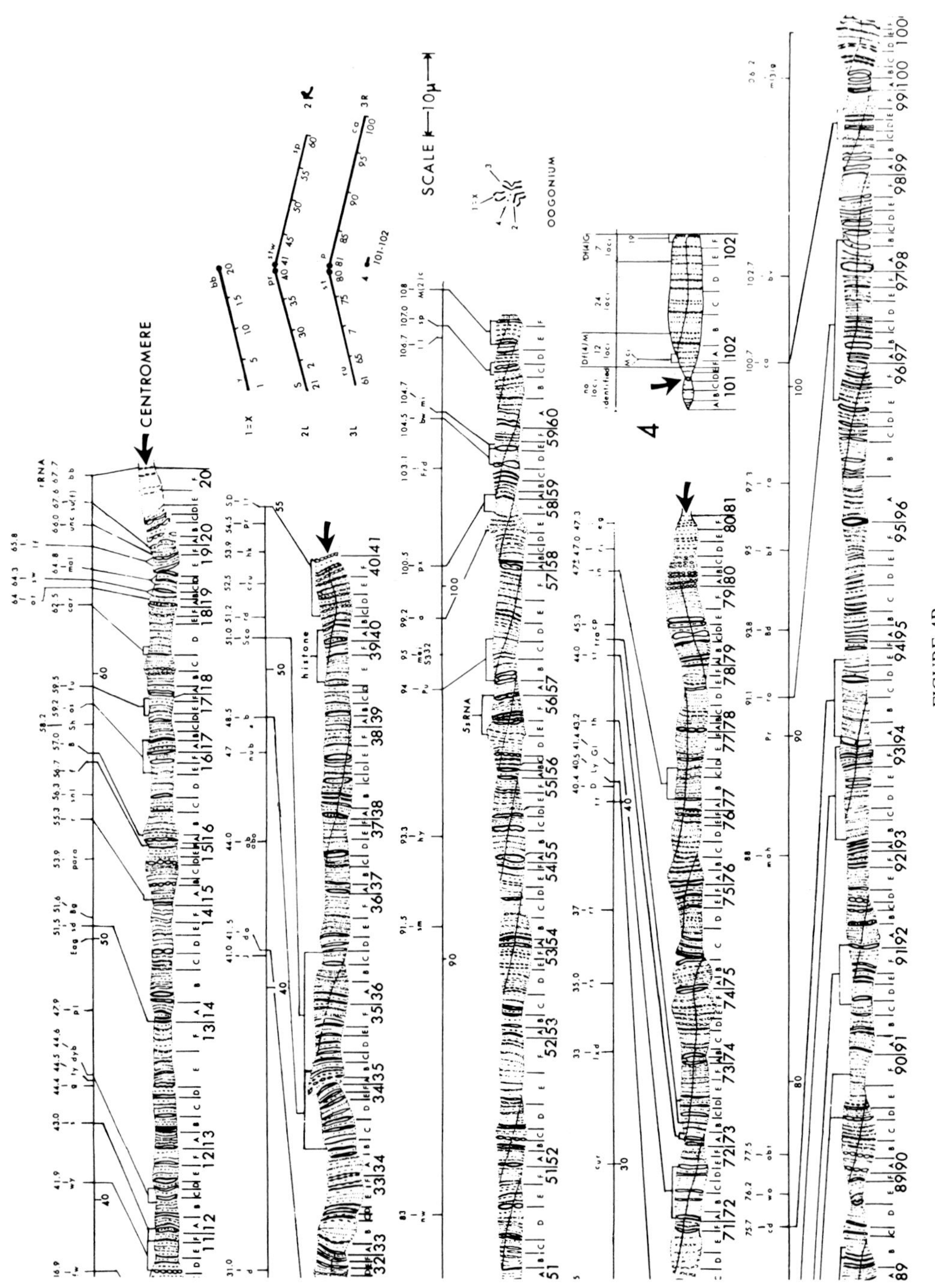

FIGURE 4B.

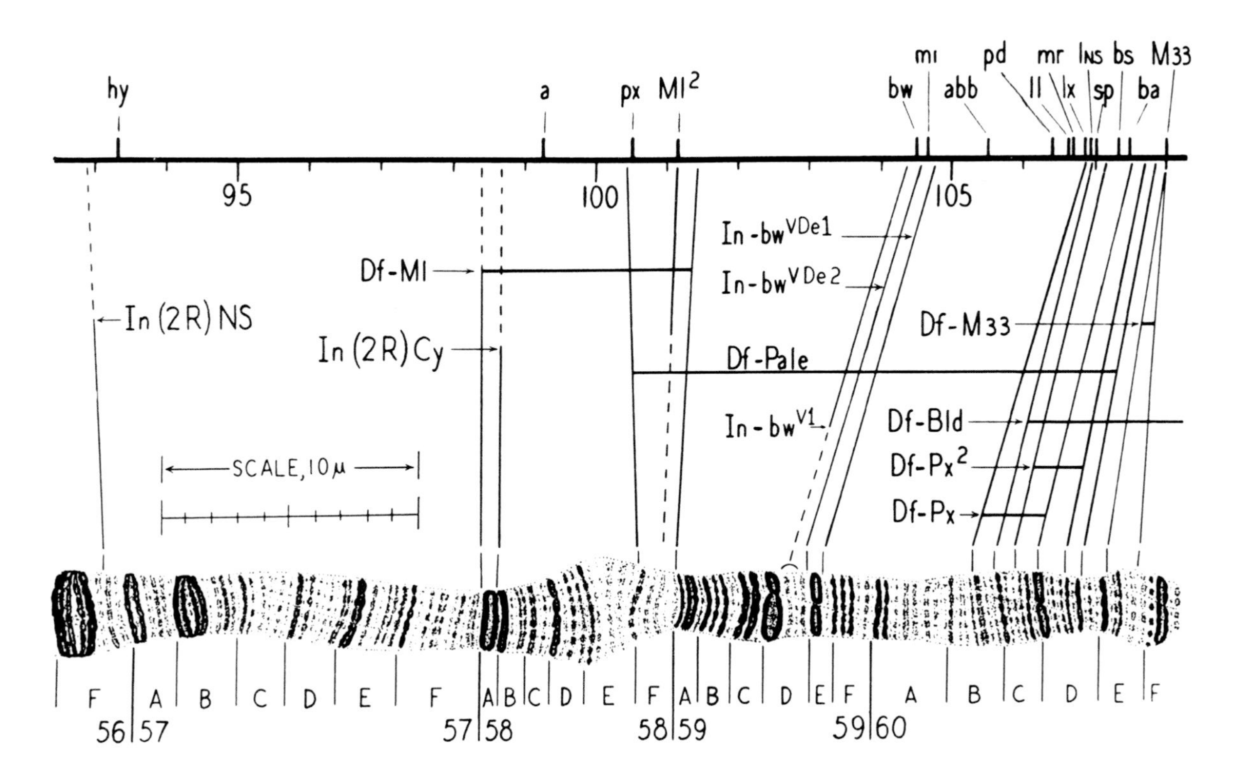

FIGURE 5. C. B. Bridges' revised map of the distal divisions of 2R chromosome, comparing the location of genes in the genetic and chromosome map. The map demonstrates also the chromosomal sites of several deficiencies and ends of inversions. (From Bridges, C. B., *Cytologia, Fujii Jub.*, 745, 1937. With permission.)

REFERENCES

1. **Hsu, T. C.,** *Human and Mammalian Cytogenetics, An Historical Perspective,* Springer-Verlag, New York, 1979.
2. **Painter, T. S.,** A cytological map of the X chromosome of *Drosophila melanogaster, Science,* 73, 647, 1933.
3. **Painter, T. S.,** A cytological map of the X-chromosome of *Drosophila melanogaster, Anat. Rec.,* 51, 111, 1931.
4. **Painter, T. S.,** A new method for the study of chromosome rearrangements and the plotting of chromosome maps, *Science,* 78, 585, 1933.
5. **Painter, T. S.,** A new method for the study of chromosome aberrations and the plotting of chromosome map in *Drosophila melanogaster, Genetics,* 19, 175, 1934.
6. **Painter, T. S.,** The morphology of the X chromosome in salivary glands of *Drosophila melanogaster,* and a new method of chromosome map for this element, *Genetics,* 19, 448, 1934.
7. **Painter, T. S.,** Salivary chromosomes and the attack on the gene, *J. Hered.,* 25, 465, 1934.
8. **Painter, T. S.,** The morphology of the third chromosome in the salivary gland of *Drosophila melanogaster,* and a new map of this element, *Genetics,* 20, 301, 1935.
9. **Heitz, E.,** Die somatische Heteropyknose bei *Drosophila melanogaster* und ihre genetische Bedeutung, *Z. Zellforsch. Mikr. Anat.,* 20, 237, 1934.
10. **Morgan, T. H.,** Personal recollections of Calvin B. Bridges, *J. Hered.,* 30, 355, 1939.
11. **Shine, I. and Wrobel, S.,** *Thomas Hunt Morgan, Pioneer of Genetics,* The University Press of Kentucky, Lexington, 1976.
12. **Bridges, C. B.,** Salivary chromosome maps with a key to the banding of the chromosomes of *Drosophila melanogaster, J. Hered.,* 26, 60, 1935.
13. **Bridges, C. B.,** A revised map of the salivary gland X-chromosome of *Drosophila melanogaster, J. Hered.,* 29, 11, 1938.
14. **Bridges, C. B. and Bridges, P. N.,** A new map of the second chromosome. A revised map of the right limb of the second chromosome of *Drosophila melanogaster, J. Hered.,* 30, 475, 1939.
15. **Bridges, P. N.,** A revised map of the left limb of the third chromosome of *Drosophila melanogaster, J. Hered.,* 32, 64, 1941.
16. **Bridges, P. N.,** A revision of the salivary gland 3R-chromosome map of *Drosophila melanogaster, J. Hered.,* 32, 299, 1941.
17. **Bridges, P. N.,** A new map of the salivary gland 2L-chromosome of *Drosophila melanogaster, J. Hered.,* 33, 403, 1942.
18. **Koltzoff, N. K.,** The structure of chromosomes in the salivary glands of *Drosophila, Science,* 80, 312, 1934.
19. **Laird, C. D.,** DNA of *Drosophila* chromosomes, *Annu. Rev. Genet.,* 7, 177, 1973.
20. **Ohno, S.,** *Evolution by Gene Duplication,* Springer-Verlag, 1970.
21. **Keyl, H.-G.,** Duplikation von Untereinheiten der chromosomalen DNS während der Evolution von *Chironomus thummi, Chromosoma,* 17, 139, 1965.
22. **Zipkas, D. and Riley, M.,** Proposal concerning mechanism of evolution of the genome of *Escherichia coli, Proc. Natl. Acad. Sci. U.S.A.,* 72, 1354, 1975.
23. **Nagl, W.,** The evolution of chromosomal DNA redundancy: ontogenetic lateral, versus phylogenetic tandem changes, *Nucleus,* 20, 10, 1977.
24. **Lefevre, G., Jr.,** A photographic representation of the polytene chromosomes of *Drosophila melanogaster* salivary glands, in *The Genetics and Biology of Drosophila,* Vol. 1a, Ashburner, M. and Novitski, E., Eds., Academic Press, New York, 1976., 31.
25. **King, T. C.,** *Genetics,* 2nd ed., Oxford University Press, London, 1965.
26. **King, R. C.,** *Ovarian Development in Drosophila melanogaster,* Academic Press, New York, 1970.
27. **Slizynski, B. M.,** A revised map of salivary gland chromosome IV, *J. Hered.,* 35, 322, 1944.
28. **Lindsley, D. L. and Grell, E. H.,** Genetic variations of *Drosophila melanogaster,* Carnegie Institution of Washington Publ. 627, 1968.
29. **King, R. C., Ed.,** *Drosophila melanogaster:* an introduction, in *Handbook of Genetics,* Vol. 3, Plenum Press, New York, 1975, 625.
30. **Darlington, C. D.,** The chromosome revolution, in *Chromosomes Today,* Vol. 6, de la Chapelle, A. and Sorsa M., Eds., Elsevier/North-Holland, Amsterdam, 1977, 1.
31. **Bridges, C. B.,** Correspondences between linkage maps and salivary chromosome structure, as illustrated in the tip of chromosome 2R of *Drosophila melanogaster, Cytologia, Fujii Jub.,* 745, 1937.
32. **Sorsa, V.,** Ultrastructure of the 5S RNA locus in the salivary gland chromosomes of *Drosophila melanogaster, Hereditas,* 74, 297, 1973.
33. **Lefevre, G., Jr. and Wilkins, M. D.,** Cytogenetic studies on the white locus in *Drosophila melanogaster, Genetics,* 53, 175, 1966.

34. **Lefevre, G., Jr. and Green, M. M.,** Genetic duplication in the *white-split* interval of the X chromosome in *Drosophila melanogaster, Chromosoma,* 36, 391, 1972.
35. **Sorsa, V., Green, M. M., and Beermann, W.,** Cytogenetic fine structure and chromosomal localization of *white* gene in *Drosophila melanogaster, Nature New Biol.,* 245, 43, 1973.
36. **Lewis, E. B.,** The *Star* and *asteroid* loci in *Drosophila melanogaster, Genetics,* 27, 153, 1942.
37. **Berger, C. A.,** The uniformity of the gene complex, *J. Hered.,* 31, 3, 1940.
38. **Pavan, C. and Breuer, M. E.,** Polytene chromosomes in different tissues of *Rhynchosciara, J. Hered.,* 43, 152, 1952.
39. **Beermann, W.,** Chromomerenkonstanz und spezifische Modifikation der Chromosomenstruktur in der Entwicklung und Organdifferenzierung von *Chironomus tentans, Chromosoma,* 5, 139, 1952.
40. **Slizynski, B. M.,** *Chironomus* versus *Drosophila, J. Genet.,* 50, 77, 1950.
41. **Beermann, W.,** Riesenchromosomen, *Protoplasmatologia,* VI, D, Springer-Verlag, Wien, 1962.
42. **Sengün, A. and Kosswig, C.,** Weiteres über den Bau der Riesenchromosomen in verschiedenen Geweben von *Chironomus*-larven, *Chromosoma,* 3, 195, 1947.
43. **Sengün, A.,** Variability of banding patterns of giant chromosomes, *J. Hered.,* 45, 119, 1954.
44. **Crick, F.,** General model for the chromosomes of higher organisms, *Nature,* 234, 25, 1971.
45. **Speiser, C.,** Eine Hypothese über die funktionelle Organization der Chromosomen der höherer Organismen, *Theor. Appl. Genet.,* 44, 97, 1974.
46. **Zhimulev, I. F. and Belyaeva, E. S.,** Proposals to the problem of structural and functional organization of polytene chromosomes, *Theor. Appl. Genet.,* 45, 335, 1975.
47. **Zhimulev, I. F., Belyaeva, E. S. and Semeshin, V. F.,** Information content of polytene chromosome bands and puffs, *CRC Crit. Rev. Biochem.,* 11, 303, 1981.
48. **Ribbert, D.,** Unterschiedliche Chromomerenmuster von Polytänchromosomen im Keimbahn und Soma der Fliege *Calliphora erythrocephala, Nachr. Acad. Wiss. Göttingen,* 11, 189, 1975.
49. **King, R. C., Riley, S. F., Cassidy, J. D., White, P. R., and Paik, Y. K.,** Giant polytene chromosomes from the ovaries of a *Drosophila* mutant, *Science,* 212, 441, 1981.
50. **Heino, T. I.,** The banding pattern of 3R polytene chromosome from the pseudo-nurse cells of *Drosophila melanogaster Otu*-mutant, *9th Eur. Dros. Res. Conf.,* Abstr. 1985.
51. **Richards, G.,** The polytene chromosomes in the fat body nuclei of *Drosophila melanogaster, Chromosoma,* 79, 241, 1980.
52. **Richards, G.,** Polytene chromosomes, in *Comprehensive Insect Physiology, Biochemistry and Pharmacology,* Vol. 2., Kerkut, G. A. and Gilbert, L. I., Eds., Pergamon Press, Oxford, 1985, 255.

Chapter 16

ELECTRON MICROSCOPIC MAPPING OF POLYTENE CHROMOSOMES

I. THIN-SECTION ELECTRON MICROSCOPY IN MAPPING OF THE POLYTENE CHROMOSOMES

Thin sectioning of unsquashed polytene nuclei rarely if ever gives longitudinal thin sections of polytene chromosomes which are long enough for identifying the regions and bands according to the reference maps compiled by using light microscopy of squashes. Recognition of only very special structures like Balbiani rings[1] has been possible from thin-sectioned nuclei prepared for the EM by using traditional procedures. Comparison of light and electron micrographs taken from the same regions of polytene chromosomes became possible by thin sectioning of smeared or squashed salivary gland chromosomes[2,3] (Volume I, Chapter 8).

Further improvements in squash-thin-sectioning methods made it easier to compare thin-sectioned regions shown in the electron micrographs with the light microscopic reference maps,[4-6] and electron microscopic mapping of the salivary gland chromosomes of *Drosophila melanogaster* could be started.[7-10] Electron microscopy of thin-sectioned squash preparations was applied also to allocate genes in the polytene chromosomes by using small aberrations induced by the irradiation. A comparative light and electron microscopic localization of *white*-gene[11] in the salivary gland chromosomes of *D. melanogaster* was started in Tübingen 1968. Modifications of squash-thin section methods have been successfully used for the electron microscopic studies on fine structure and analyses of banding pattern of specific regions,[12-17] as well as for the more systematic mapping of whole salivary gland chromosomes.[18-29] Thin-section electron microscopy of squashed polytene chromosomes has also been applied to autoradiographic studies[30,31] of labelled nuclei, and to the photodensitometric studies on chromatin from the electron microscopic negatives.[32,33]

Thin sectioning methods of squashed chromosomes apparently offers the best resolution for high magnification electron micrographs of specific regions or bands. For the routine work the method is laborous and needs a considerable amount of patience and training especially by trimming the chromosomes and by making the axial thin sections through the whole chromosomes.

The revised map of the salivary gland chromosomes of *D. melanogaster* reconstructed on the basis of analyses of the electron micrographs taken from the thin-sectioned squash preparations is shown in Figure 1.*

II. WHOLE-MOUNT SQUASHES IN MAPPING OF POLYTENE CHROMOSOMES

Whole-mount electron microscopy of squashed polytene chromosomes has been mainly used for the structural studies by means of the high voltage electron microscopy (HVEM)[34] (see Volume I, Chapter 8). Modifications of this method, however, have been developed suitable for inspection of *in situ* hybridization both by means of the scanning electron microscopy (SEM)[35] and by means of normal transmission electron microscopy (TEM).[36] In the hybridization experiments, certainly, an accurate identification of chromosome regions according to the reference maps are necessary. Recently, the TEM of whole-mounted squashes has been successfully applied also for the mapping of the salivary gland chromosomes of *D. melanogaster*.[37-39]

* Figures 1 and 2 appear at the end of the text. Table 1 appears after the figures.

III. WHOLE-MOUNT SPREADING METHODS IN THE ELECTRON MICROSCOPIC MAPPING OF POLYTENE CHROMOSOMES

Previously, the surface spreadings of unfixed polytene chromosomes have been used mainly for ultrastructural studies (see Volume I, Chapter 8). However, the methods for using the surface spreadings of acid-treated salivary gland chromosomes in the electron microscopic studies[40] opened a possibility to identify chromosome regions and compare them with the light microscopic reference maps.[41] These methods, also called acid spreading electron microscopy (ASEM), have been developed further and used for the mapping of the salivary gland chromosomes of *D. melanogaster*.[42,43] Accordingly, also other modifications of surface-spreading procedures for the electron microscopy of acid-treated polytene chromosomes have been developed.[44,45] The method, called surface spread polytene chromosome technique (SSP),[46] has also proved to be particularly suitable for the mapping of salivary gland chromosomes.[47] Some modifications of isolating polytene chromosomes in acetic acid seem to preserve nucleosomal structure in chromatin fibers.[48]

The whole-mount methods saving the whole material of polytene chromosomes on grids can be used for quantitative analyses of chromatin material in bands and interbands. The methods can be combined to digitalized analysis of chromosomes from the electron micrographs and used for computerized mapping of chromosomes. By measuring certain parameters like diameter and thickness of bands and the length of interbands from selected electron micrographs, the banding pattern of polytene chromosomes can be plotted with computer. Even the size and form of puffed regions, continuous of dotted shape of bands, as well as possible curved shape of bands can be depicted by using the special programs[49-51] (see Figure 2).

IV. DNA CONTENT OF CHROMATIDS IN POLYTENE CHROMOSOMES

DNA contents of divisions, subdivisions, and even of individual bands of polytene chromosomes can be roughly estimated from the thickness of bands according to the following presumptions: (1) average thickness of a band corresponds to the average axial length of parallel homologous chromomeres forming the band and (2) the average thickness of all bands represents the axial length of chromomeres having an average DNA content. DNA content for an average size of interchromomere plus chromomere unit in the salivary gland chromosomes of *D. melanogaster* has been estimated to be about 21.6 kb (see Volume I, Chapter 4).

By using this average value of DNA content per average size of interchromomere-chromomere (i-c.) units the DNA contents of whole chromatids of polytene chromosomes can be calculated from the total length of bands. By summarizing the average thicknesses of all bands in a chromosome or in a part of it, and by dividing this value with the number of bands in the same region we can get average thickness of bands in this chromosome area. By dividing this value with the average chromomere length of whole genome, (which corresponds to the DNA content of 21.6 kb per i-c. unit) we can get relative length of bands in the region. Here this value is used as a length coefficient (LC) for the following simple formula for estimating DNA content per chromatid:

$$\text{LC} * \text{NUMBER OF BANDS} * 21.6\ \text{KB} = \text{DNA CONTENT} \quad (1)$$

The average thickness of bands in the whole chromosome complement of the salivary gland cells of *D. melanogaster* is about 105 nm in the AM (acetomethanol)-fixed material and about 95 nm in FAR (formaldehyde)-fixed material. The difference in the average thickness of bands in the polytene chromosomes fixed with AM and FAR is apparently caused by the fact that, particularly, the contraction of small chromomeres (narrow bands) is remarkably higher after cross-linking fixations, i.e. the axial length of small chromomeres is more increased by squashing in AM-fixed salivary gland chromosomes.

Respectively, DNA contents of i-c. units in single chromatids can be calculated per divisions and subdivisions of the salivary gland chromosomes. However, it should be pointed out that the DNA values per shorter regions like subdivisions and divisions are certainly less accurate due to differential condensation of chromatin in bands of different size. DNA content of subdivisions and divisions composed of bands mainly smaller than average size is apparently overestimated, because the small chromomeres are more stretched by squashing, especially after acetic fixatives. Correspondingly, the DNA content of subdivisions and divisions composed mainly of heavy bands may be underestimated because of the high density of chromatin in large chromomeres, particularly after cross-linking fixatives.

In the long regions like in whole arms of chromosomes the density differences of chromatin between the tightly packed heavy bands and less-condensed light bands are compensated by each other. In shorter regions the density differences of bands should be taken in account. Particularly, by estimation of DNA content of chromomeres of individual bands the densitometric measurements of relative density of chromatin are necessary for correcting the results obtained by using the average axial length of bands only. Thus the following formula should be used for estimating DNA content of chromomeres in single bands of polytene chromosomes (ADB = average density of band, ADAB = average density of average band).

$$\frac{\text{ADB}}{\text{ADAB}} * \text{LC} * 21.6 \text{ kb} = \text{DNA content per chromomere} \qquad (2)$$

Because the comparative densitometric results from all of the bands of salivary gland chromosomes are not yet available the DNA contents given in Table 1 per subdivisions and divisions are calculated by using only the axial lengths of chromomeres (Equation 1). Thus the table gives only an approximate distribution of DNA per chromatid in the subdivisions and divisions of the salivary gland chromosomes in *D. melanogaster*.

The density of chromatin in the bands of polytene chromosomes seems to be positively correlated with the axial length (the thickness) of bands, i.e. the thicker band the tighter structure. Based on this correlation a "tentative density factor" can be approximated just by comparing the total axial length of bands per subdivision or division with the corresponding average length, i.e. with the total length of same number of average bands. As mentioned above, the average axial length of bands (or chromomeres) is about 95 nm in the FAR-fixed material used mainly for compiling the Table 1. The "tentative density factor" (TDF) approximated by using the relative thickness of bands cannot replace the density measurements, but anyway, it gives an approximate how much and to which direction the DNA values determined by using the axial lengths only should be corrected. It should be pointed out that TDFs calculated for very short regions, composed of heavy bands only, may be inaccurate, giving too high "density" values. In densitometric studies of thin-sectioned band chromatin the density values are usually varying between about 0.5 and 1.5, if average density is designated by 1. In Table 1 the TDF has been calculated only for the divisions. By using Formula 3 (below) TDF can be easily calculated for subdivisions, too, by using the band number and the DNA content given for each subdivision.

$$\frac{\begin{array}{c}\text{DNA CONTENT OF SUBDIVISION (in kb)}\\ \text{(=DNA content according to band thickness)}\end{array}}{\begin{array}{c}\text{BAND NUMBER OF SUBDIVISION} \times 21.6 \text{ kb}\\ \text{(=average DNA content of bands)}\end{array}} = \text{TDF} \qquad (3)$$

By multiplying the DNA content given in the Table for the subdivision in question with the TDF calculated as shown above, a tentatively "density corrected" value for the DNA content can be estimated.

The estimates of the average axial lengths of bands (chromomeres) used for drawing the division maps (Figure 1), as well as for calculating the DNA values in Table 1 were made by using millimeter scale from electron micrographs copied with final magnifications of about 25,000 to 250,000. The maximum resolution in the original EM map drawn at magnification of × 25,000 is about 0.5 mm, corresponding to about 20 nm in thin sections, and to about 4.6 kb of DNA in chromatin of average band. Because minor errors in estimation of average thicknesses of bands evidently exist due to the variation between different preparations of squashed salivary gland chromosomes and to the different magnifications of electron micrographs in large material, the DNA values given in Table 1 should be considered very tentative ones.

FIGURE 1. The revised reference maps, shown from pages 33 to 81 as division maps, are reprinted by permission from the reports of C. B. Bridges,[13] C. B. and P. N. Bridges,[14] and P. N. Bridges,[15-17] and combined with photographic maps of T. I. Heino (Courtesy of T. I. Heino), and the electron microscopic map drawn by the author on the basis of EM photo maps compiled by A. O. Saura and T. I. Heino of a large number of electron micrographs taken from thin-sectioned salivary gland chromosomes of *Drosophila melanogaster*.

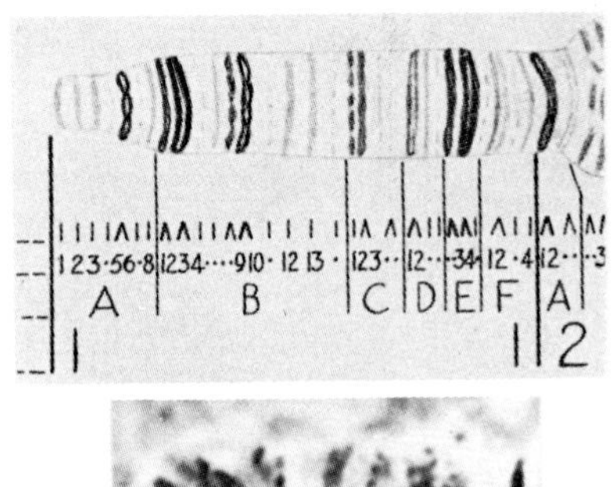

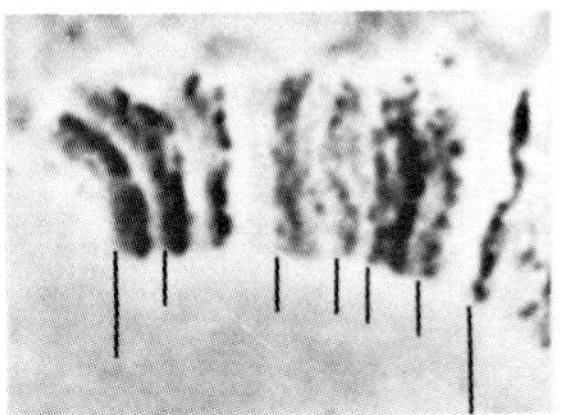

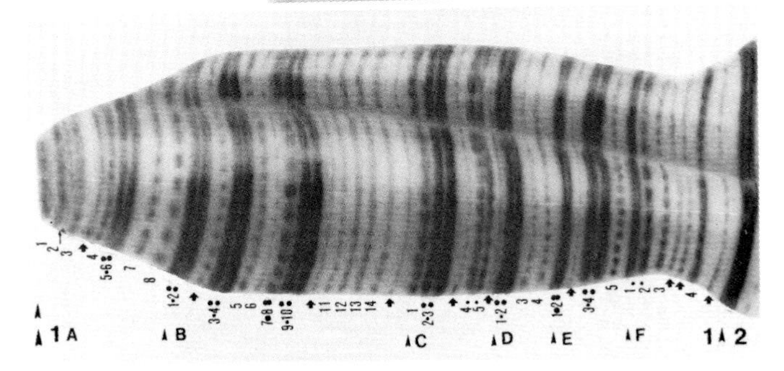

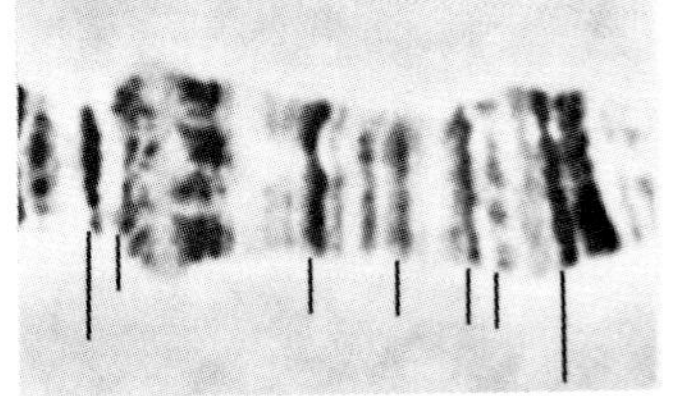

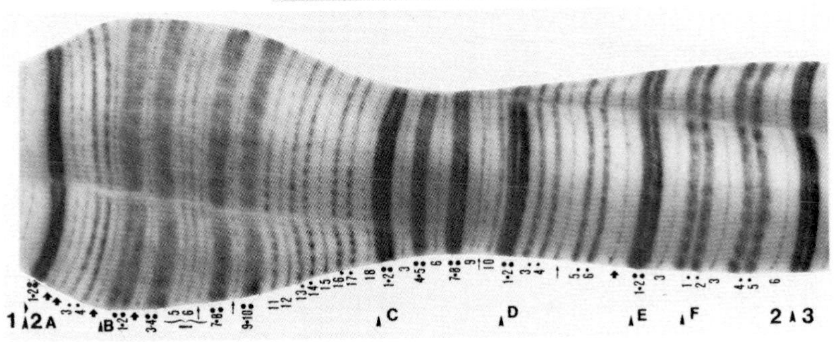

FIGURE 1 (continued)

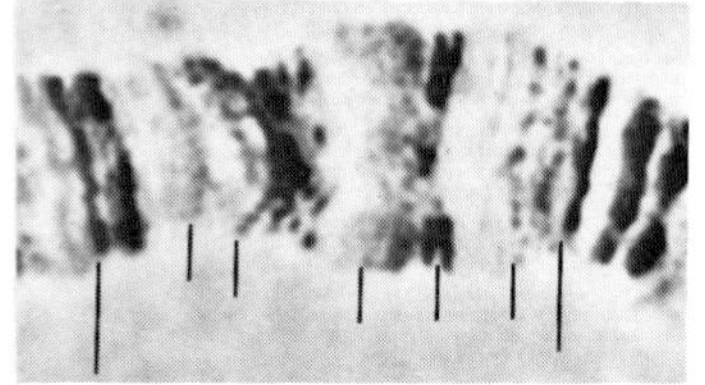

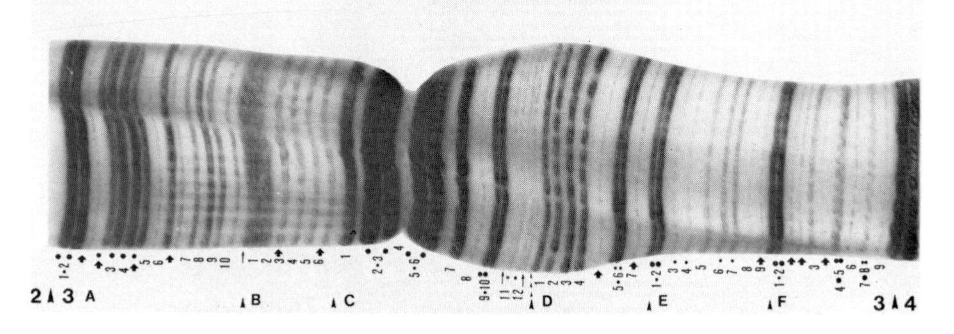

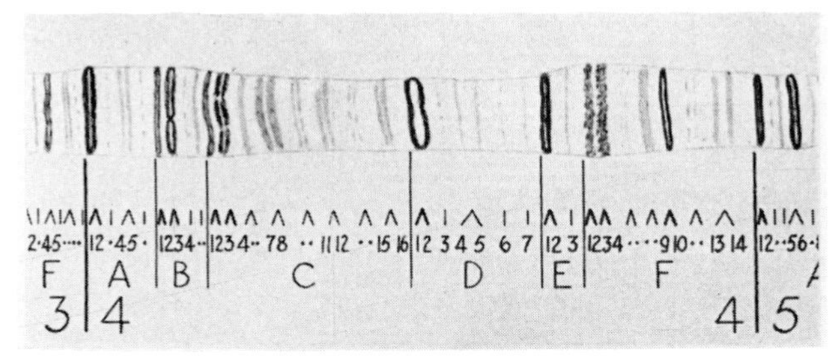

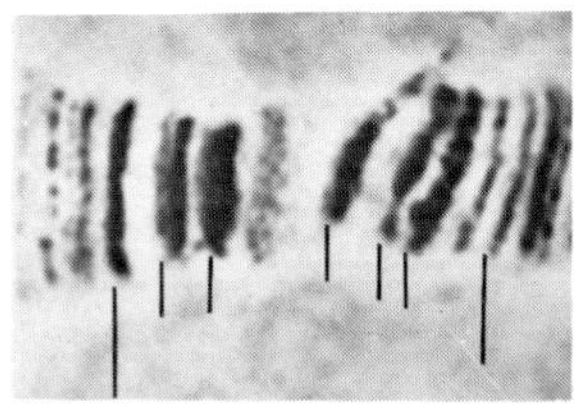

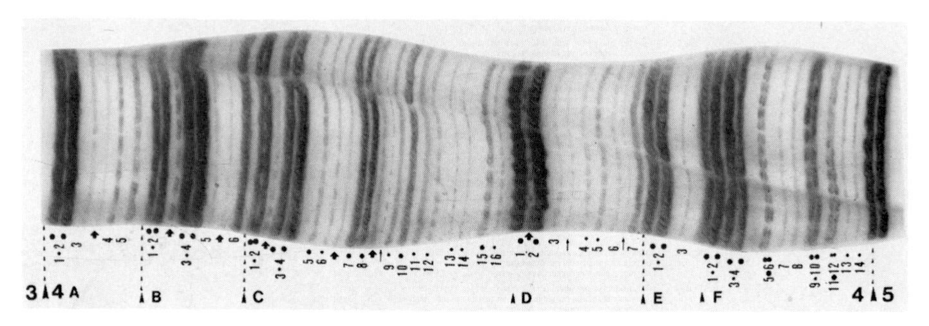

FIGURE 1 (continued)

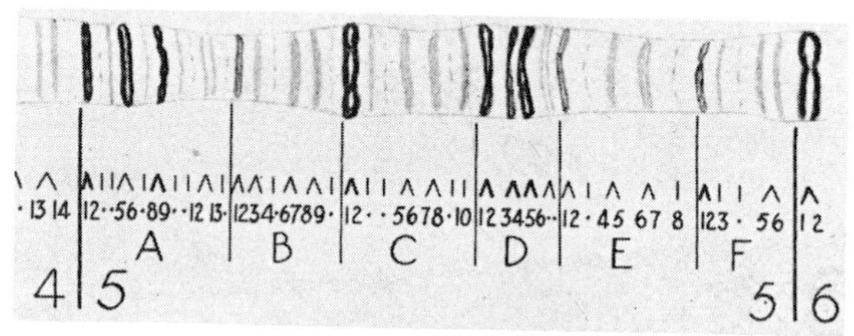

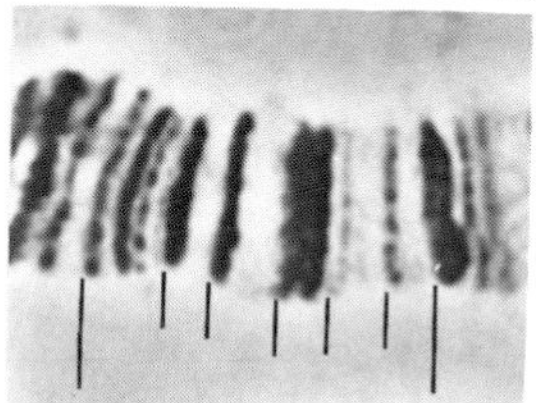

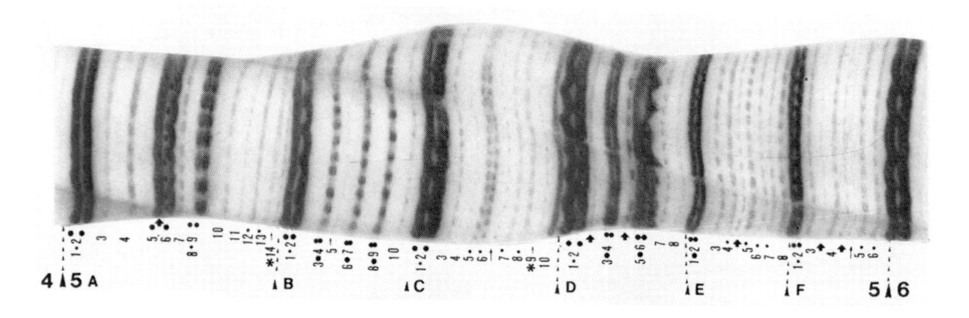

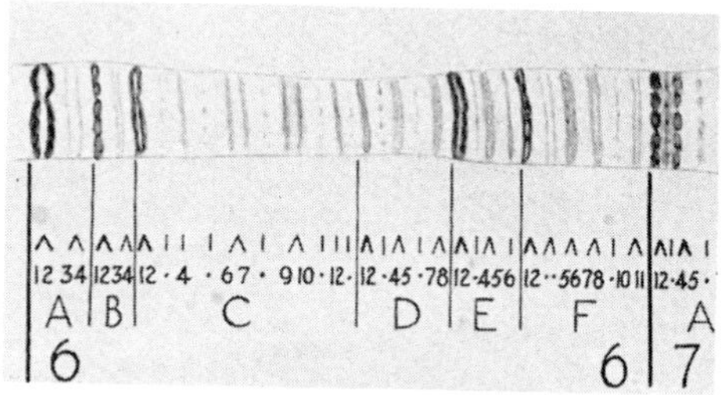

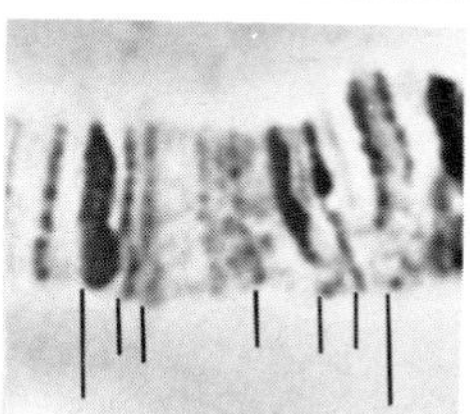

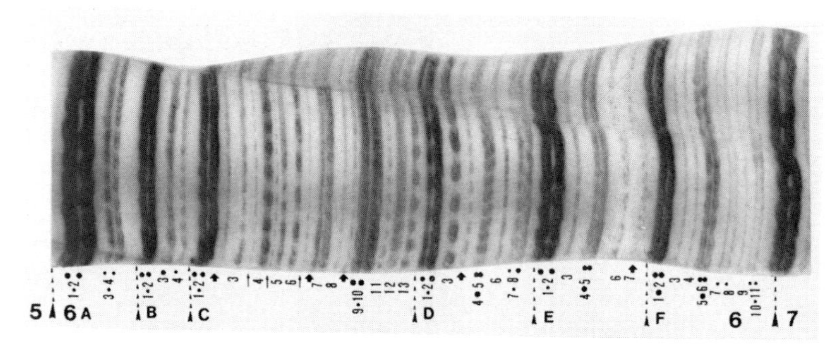

FIGURE 1 (continued)

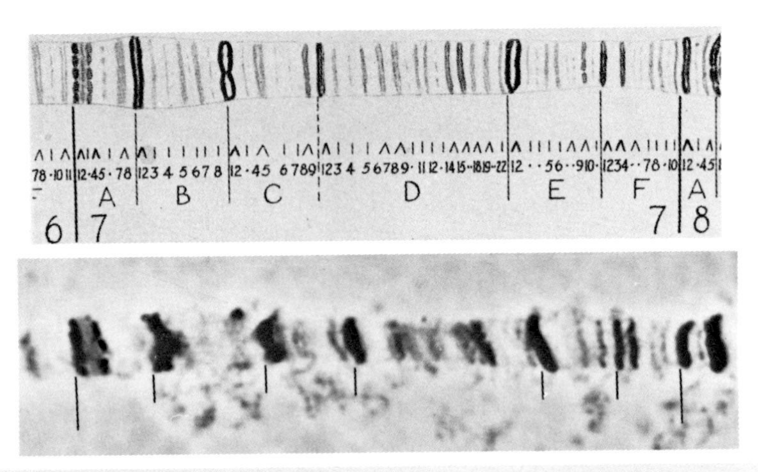

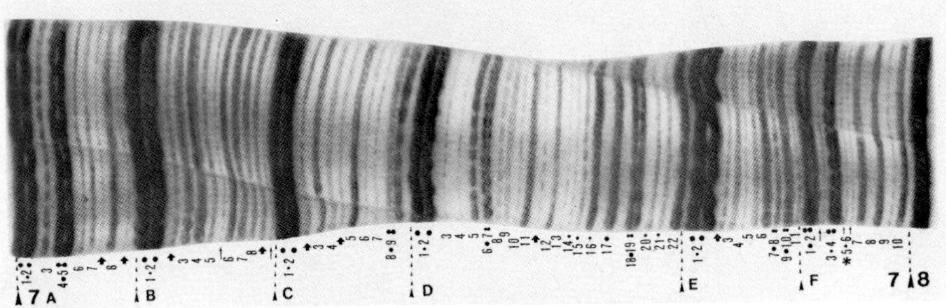

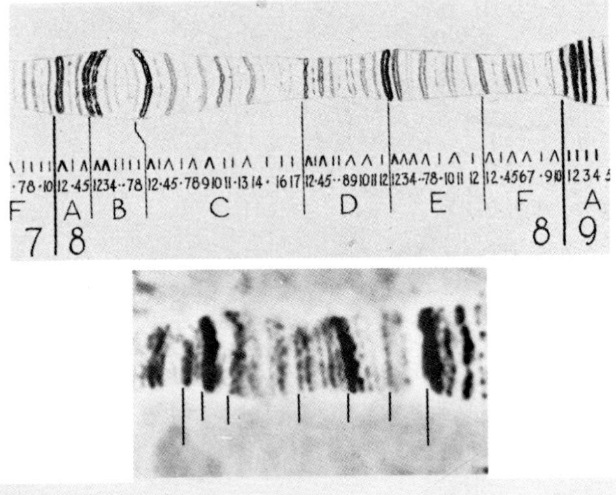

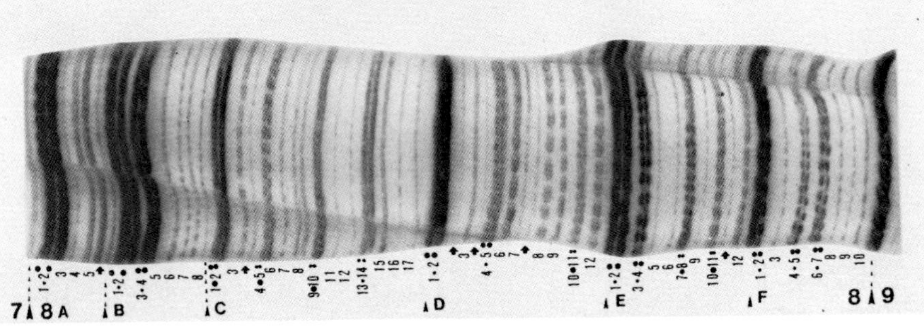

FIGURE 1 (continued)

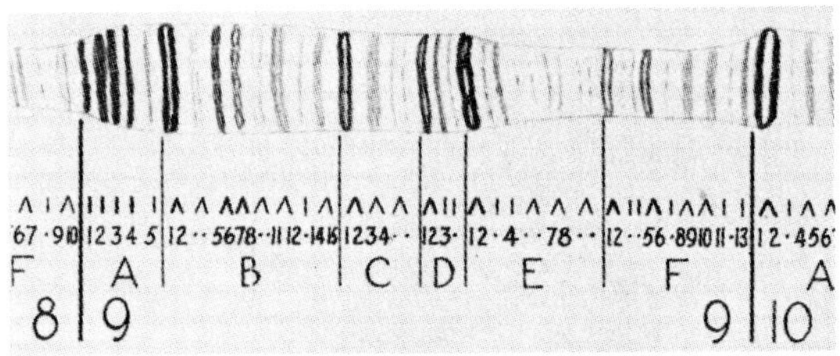

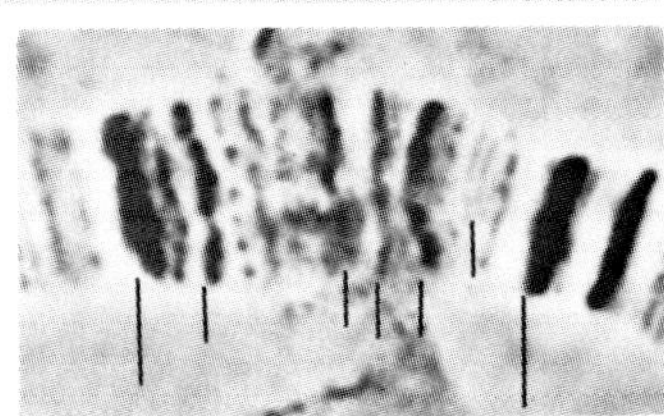

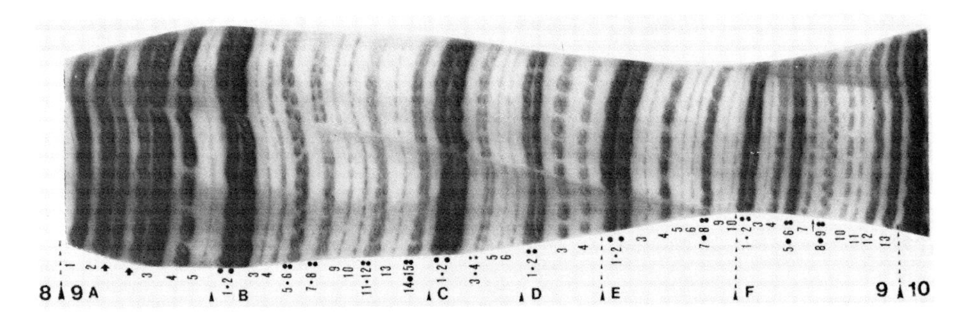

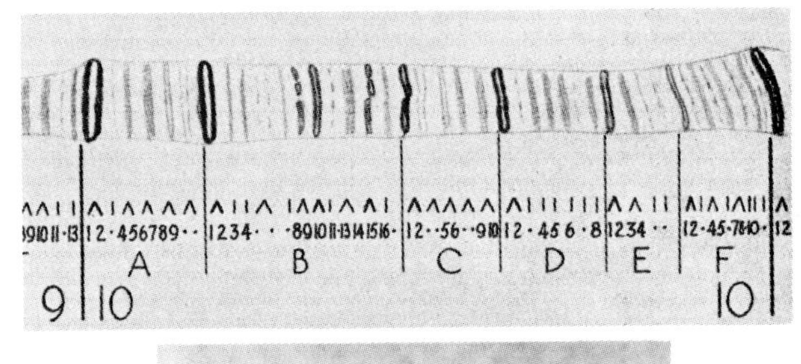

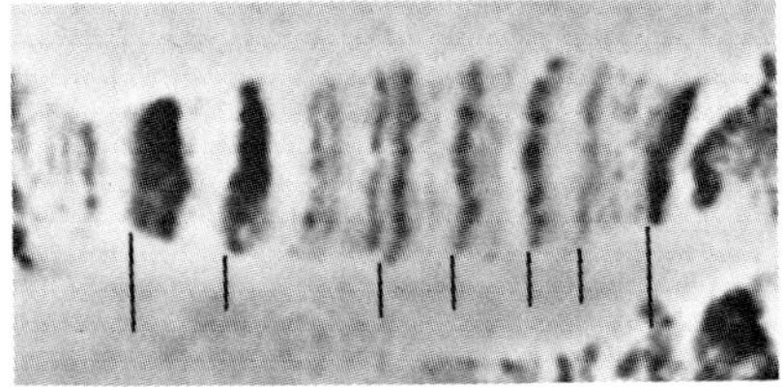

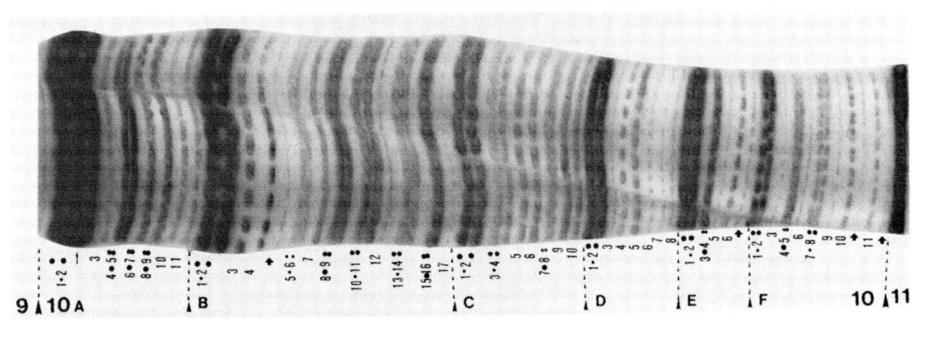

FIGURE 1 (continued)

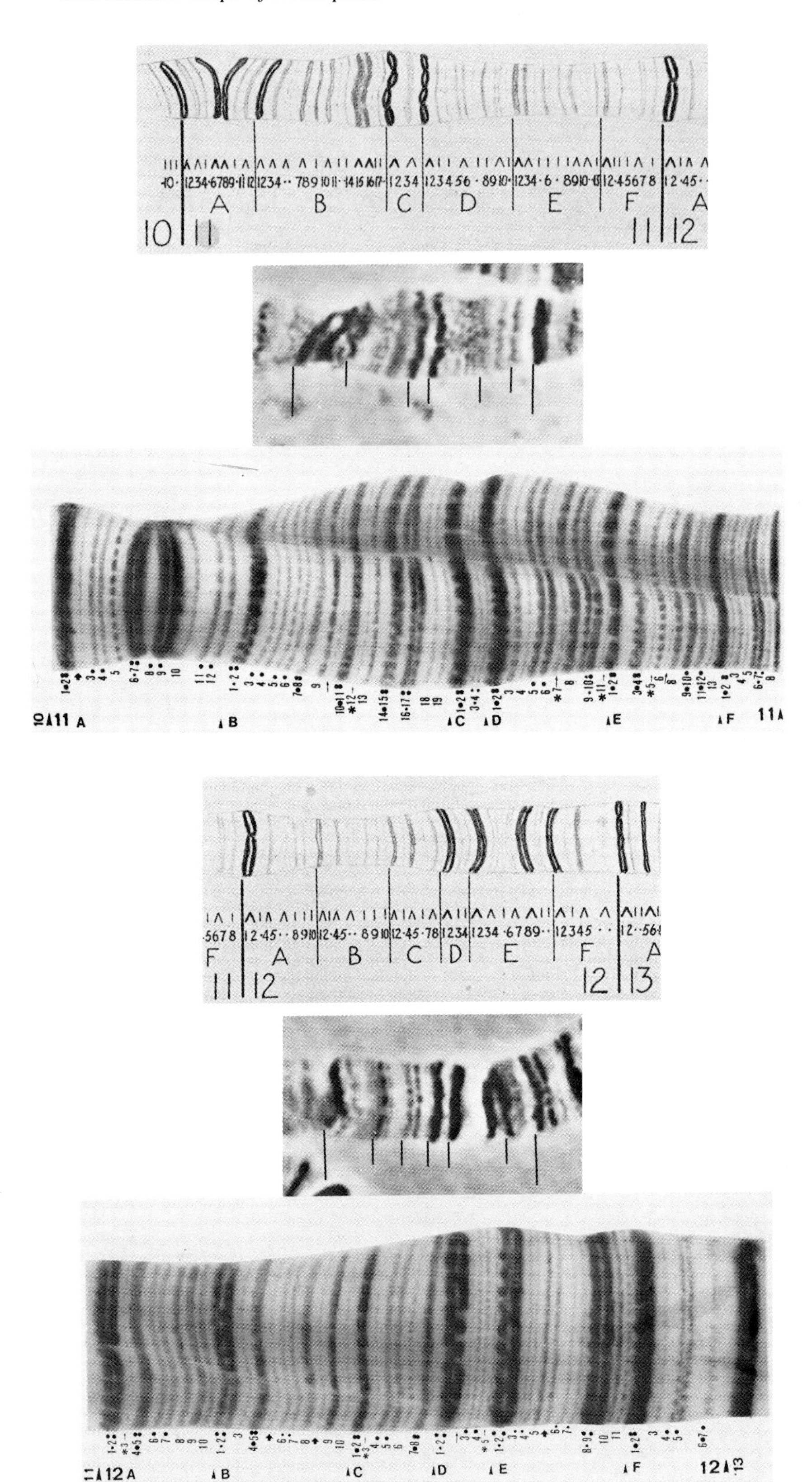

FIGURE 1 (continued)

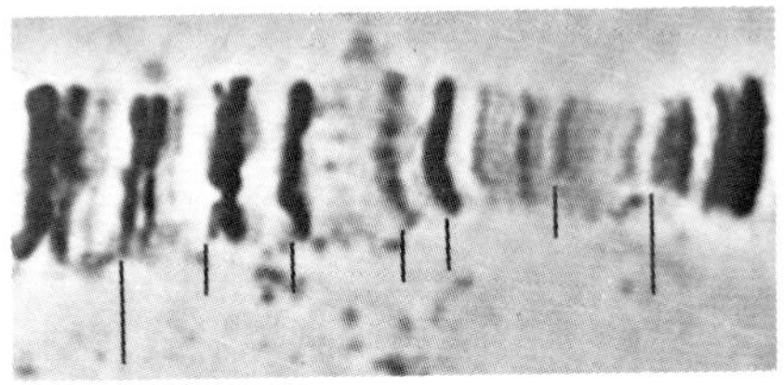

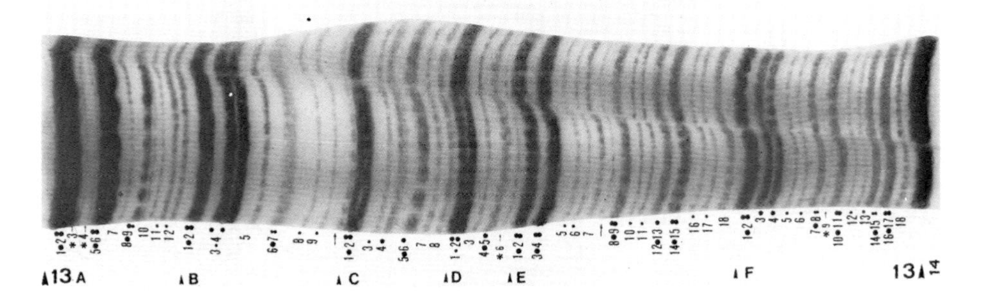

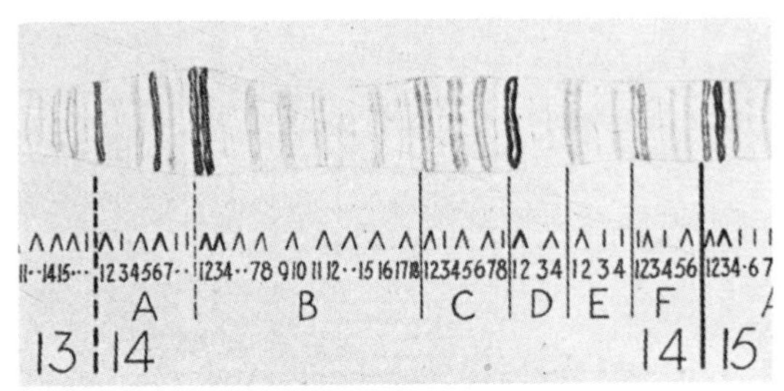

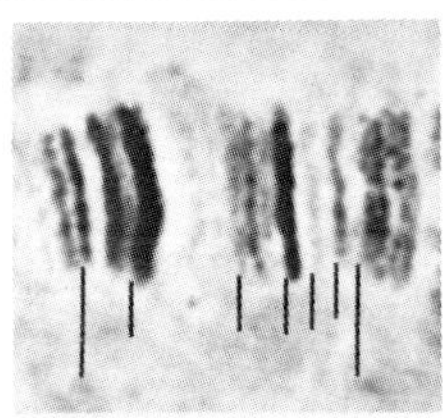

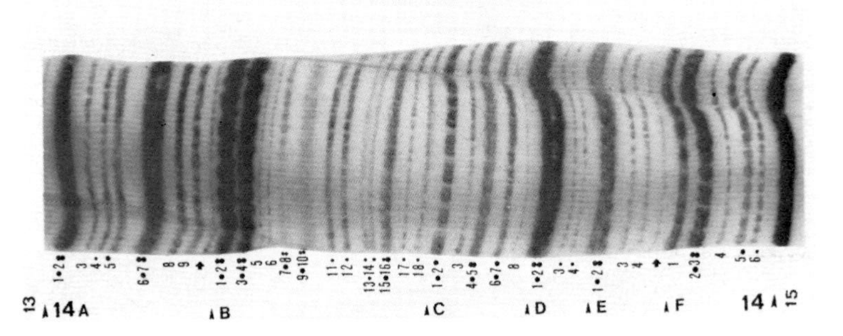

FIGURE 1 (continued)

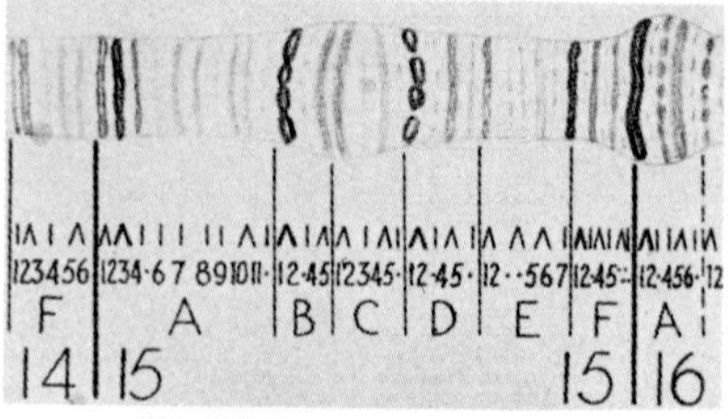

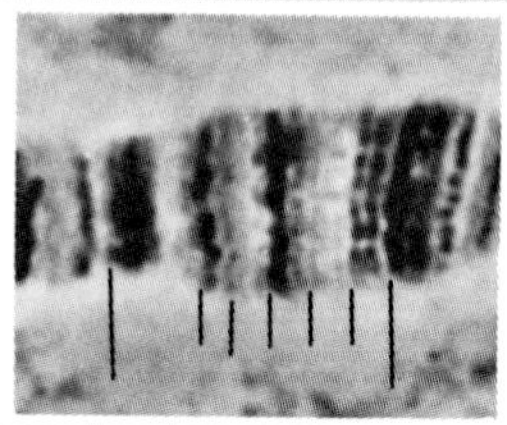

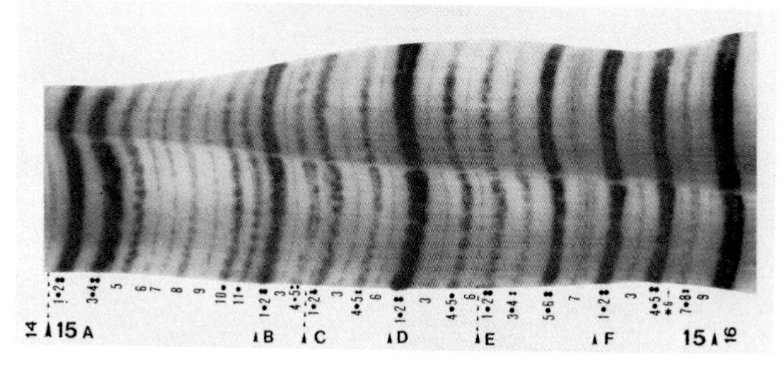

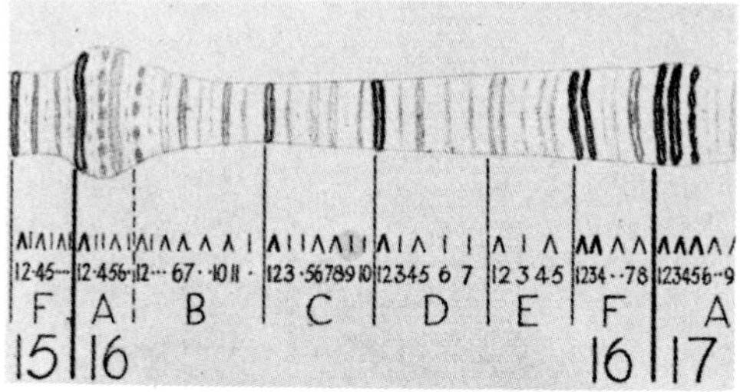

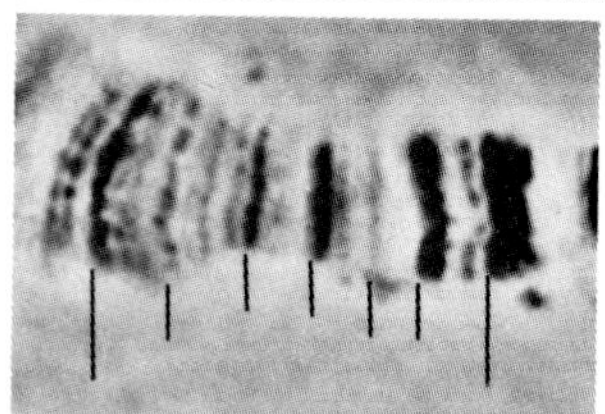

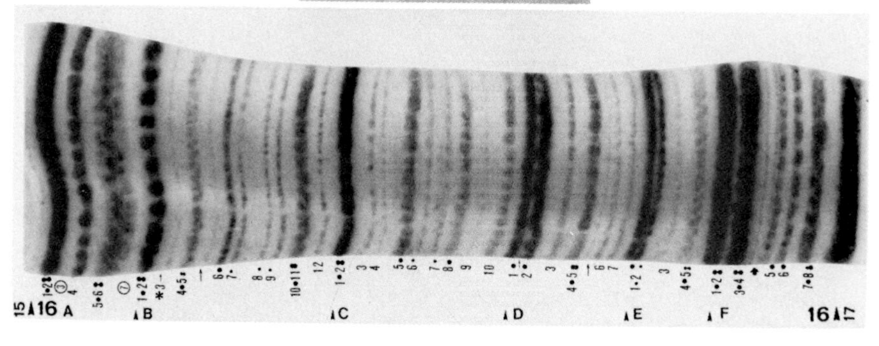

FIGURE 1 (continued)

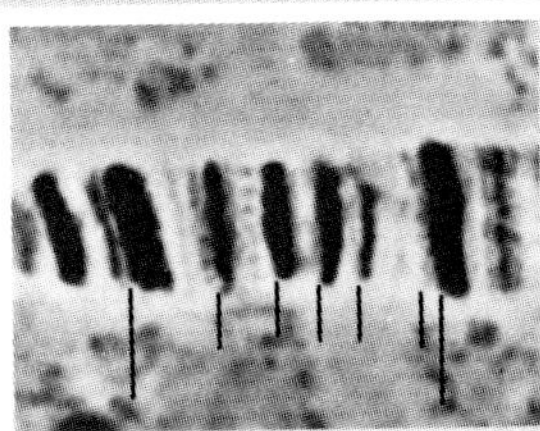

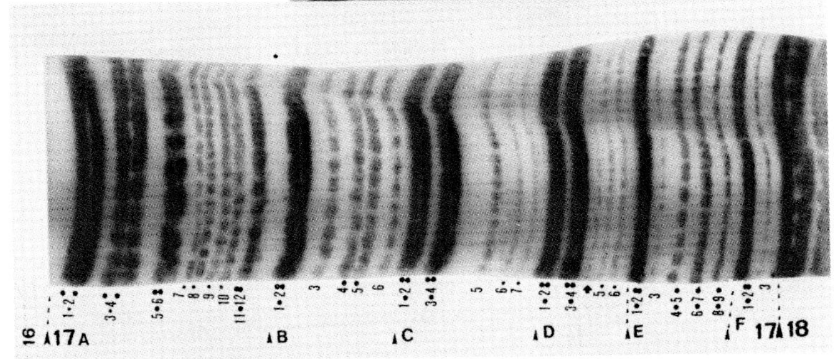

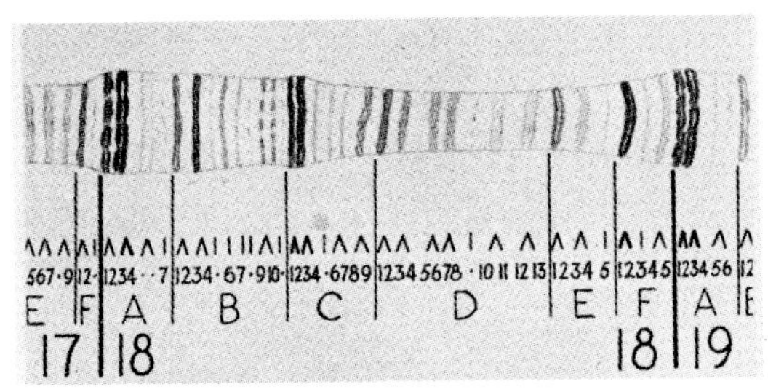

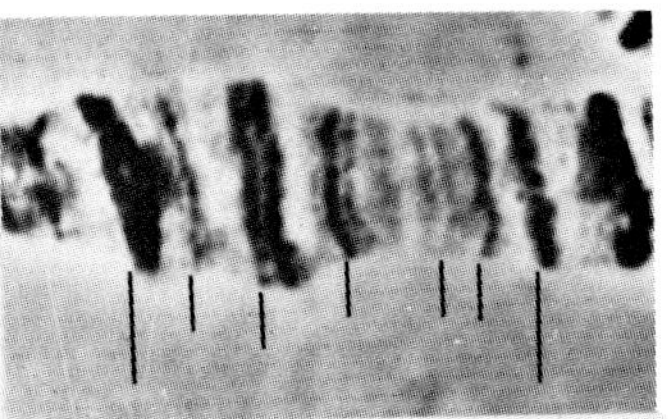

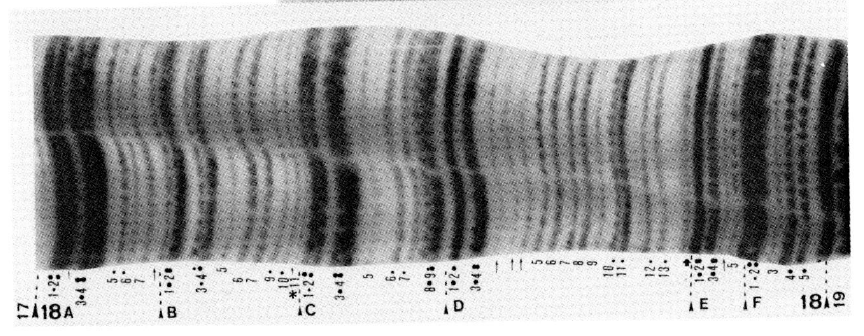

FIGURE 1 (continued)

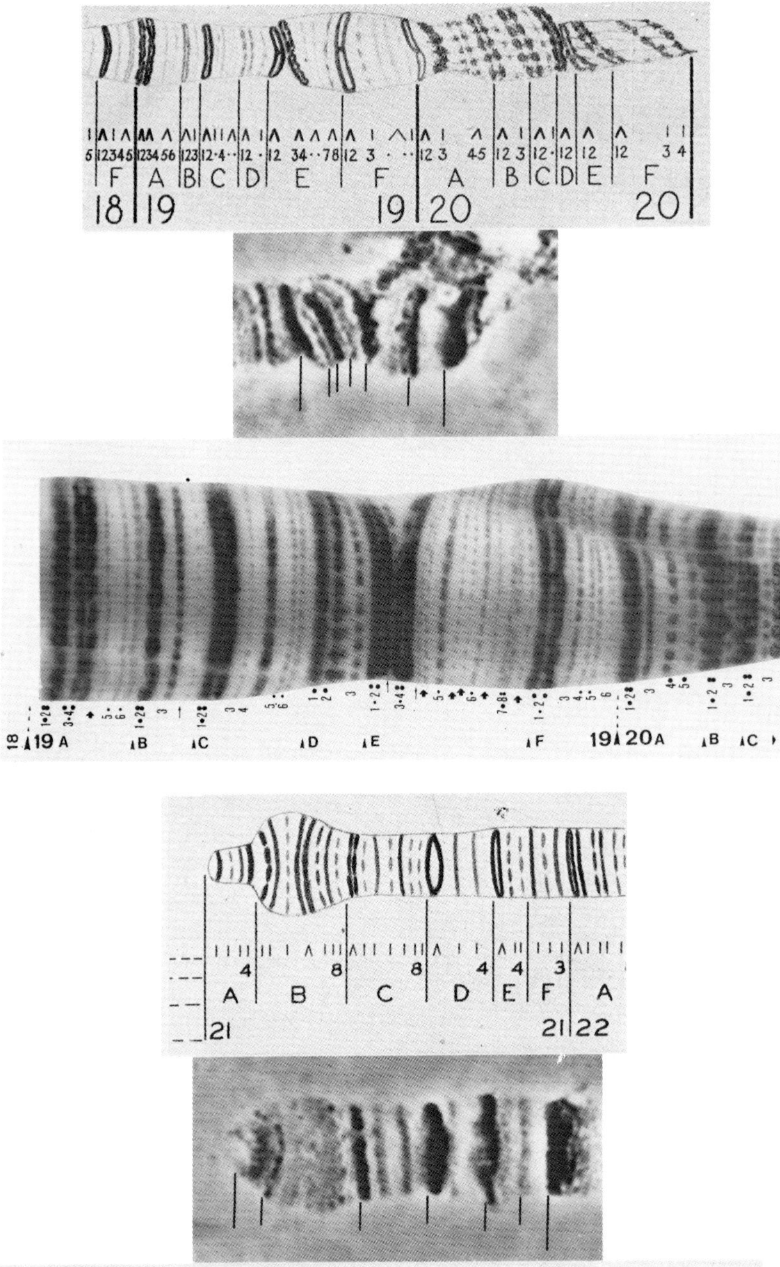

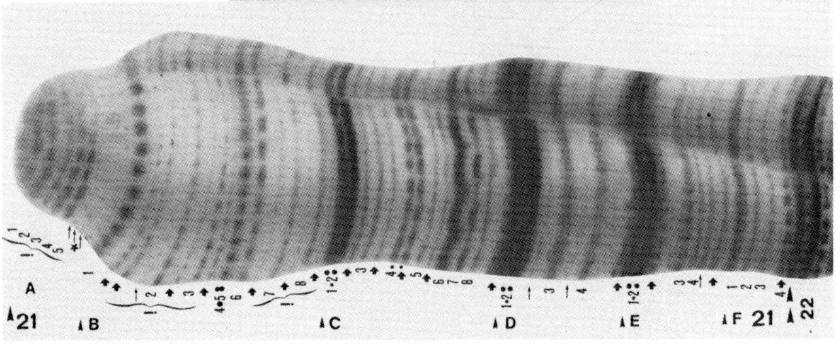

FIGURE 1 (continued)

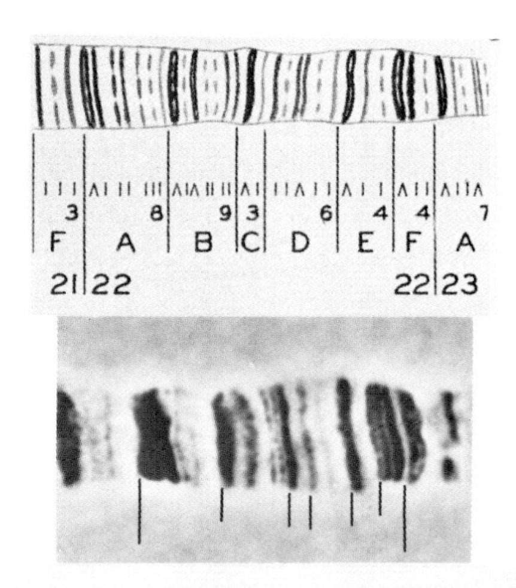

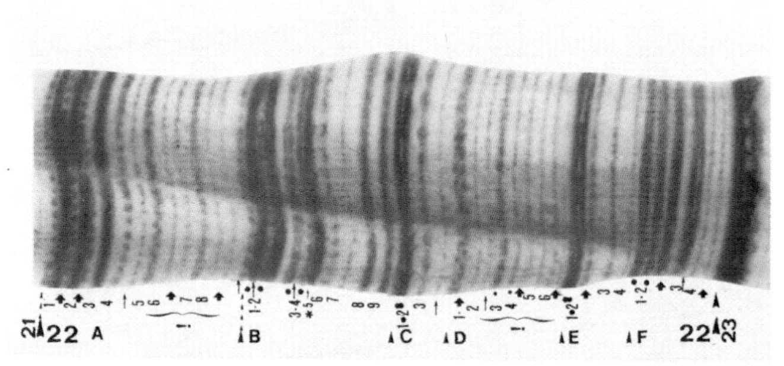

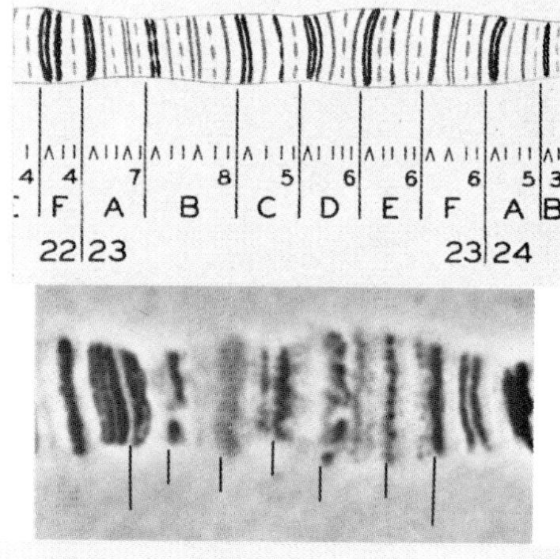

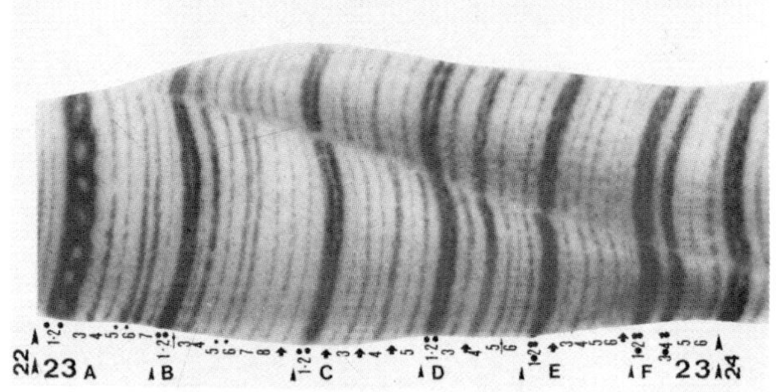

FIGURE 1 (continued)

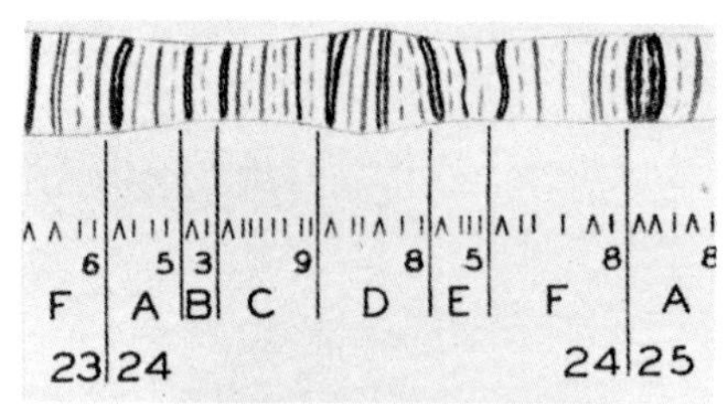

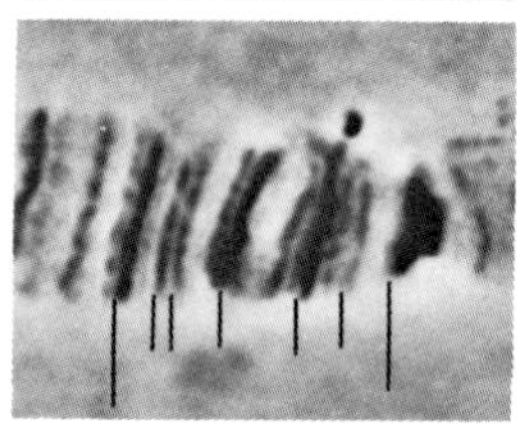

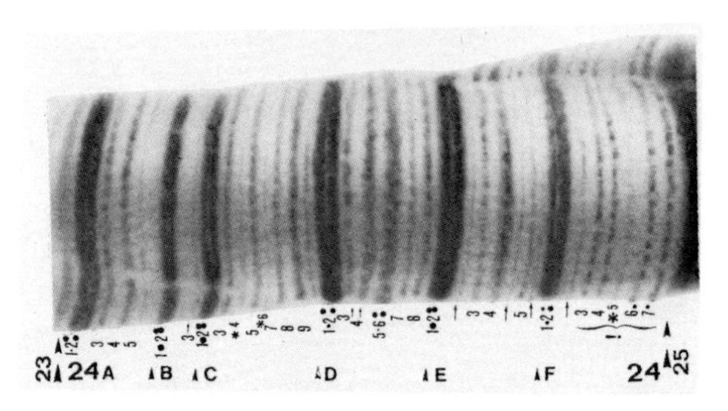

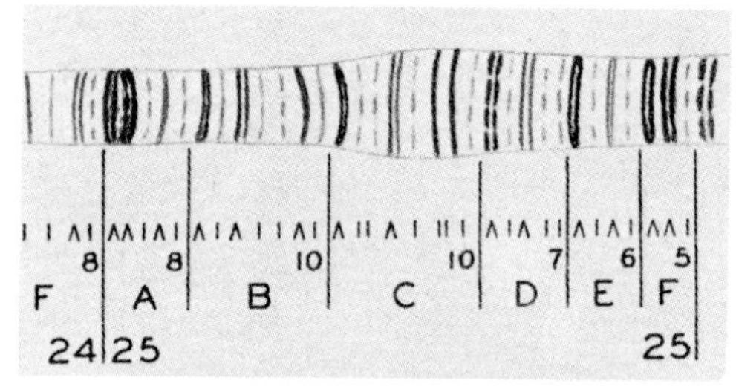

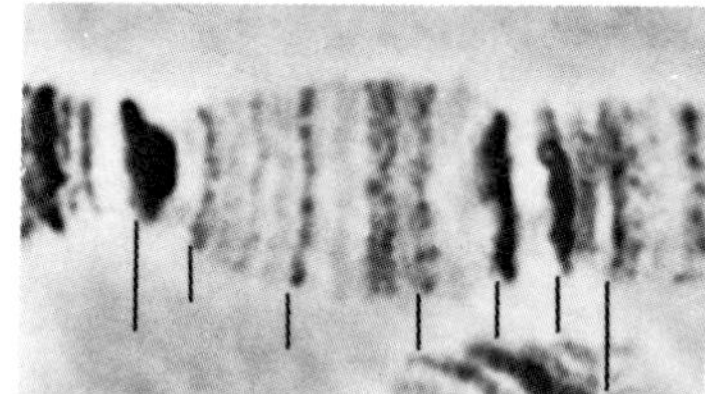

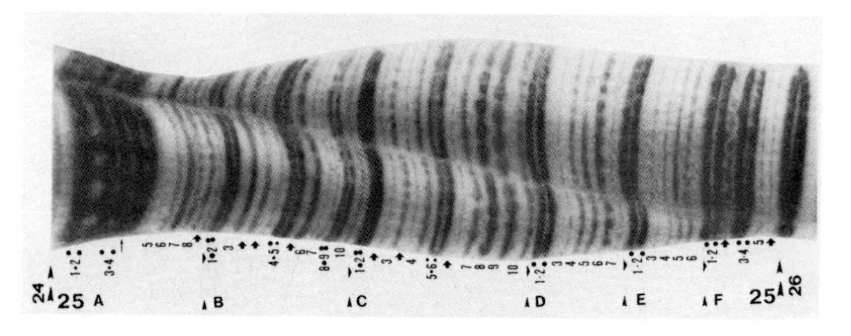

FIGURE 1 (continued)

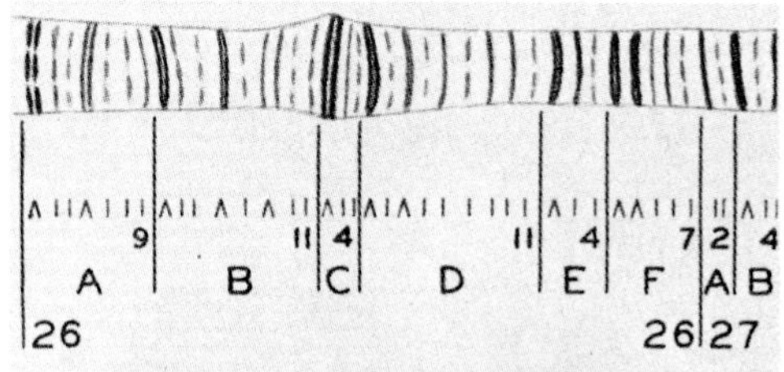

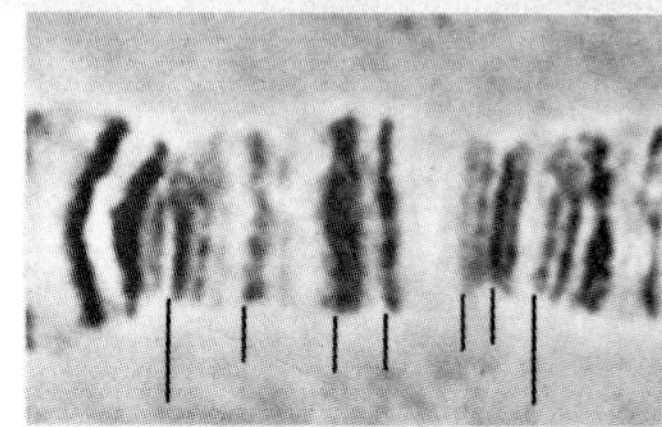

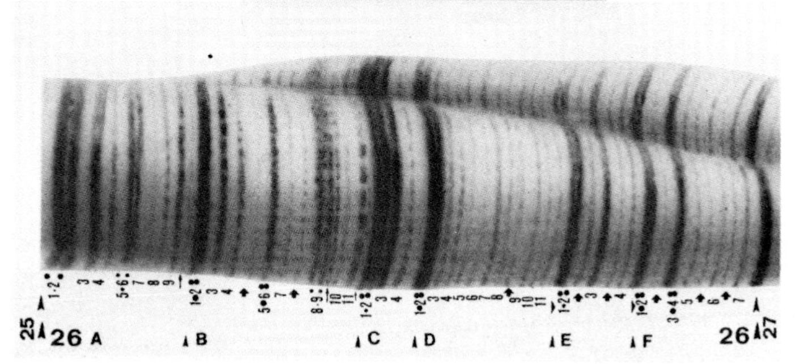

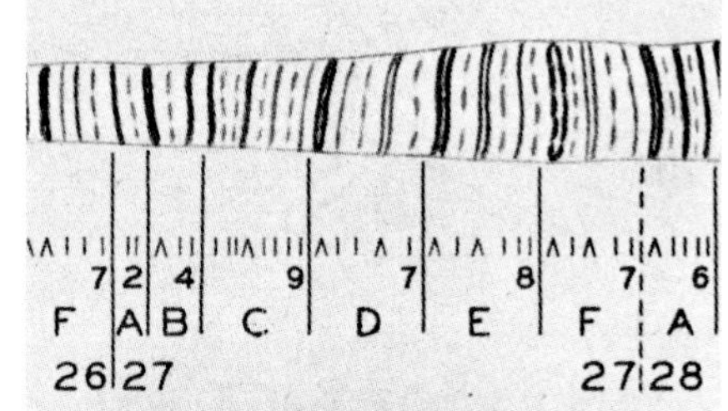

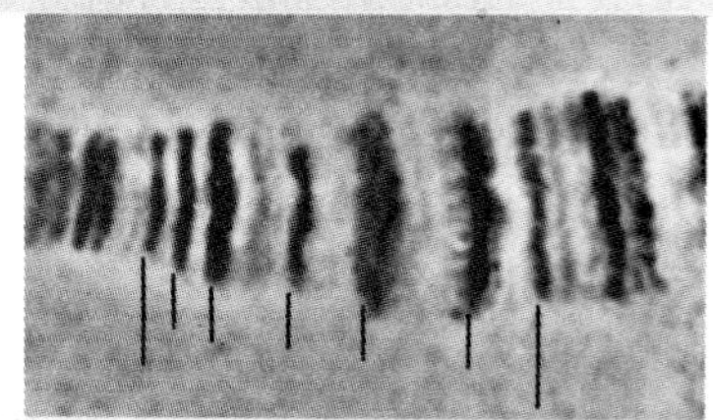

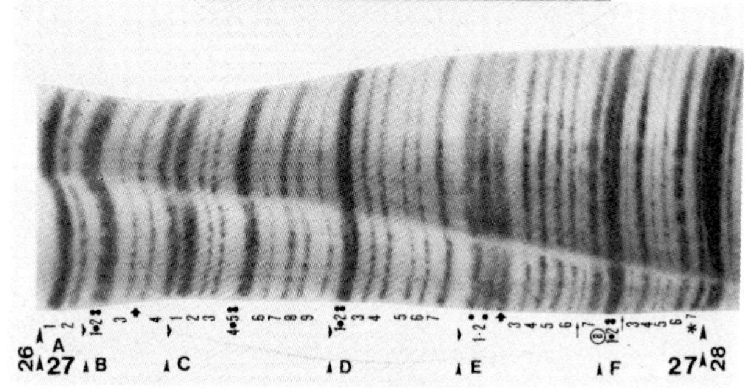

FIGURE 1 (continued)

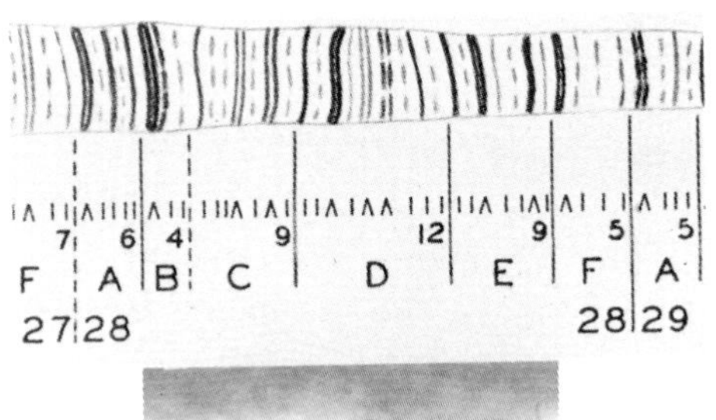

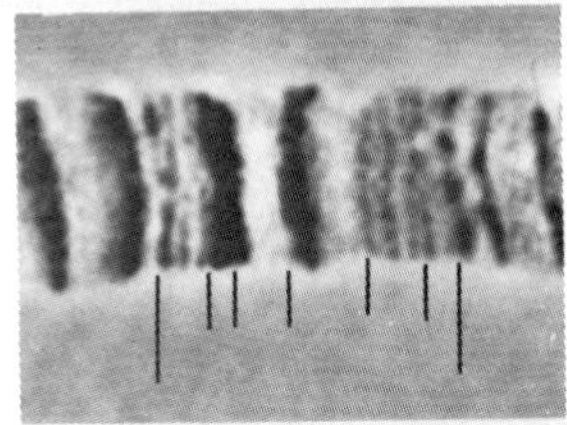

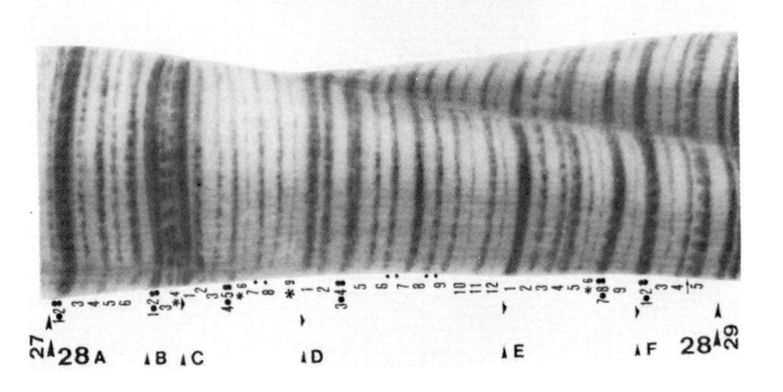

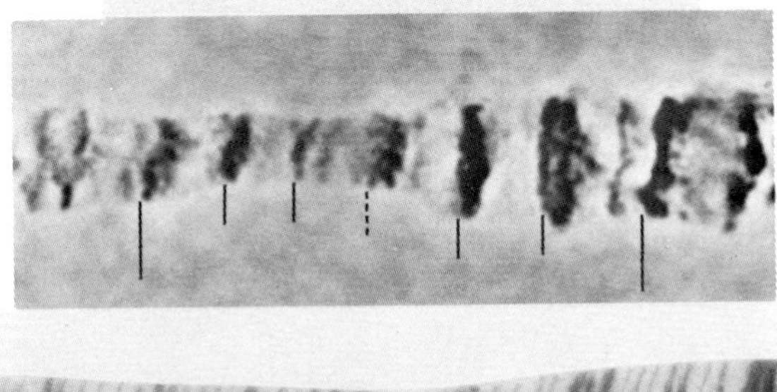

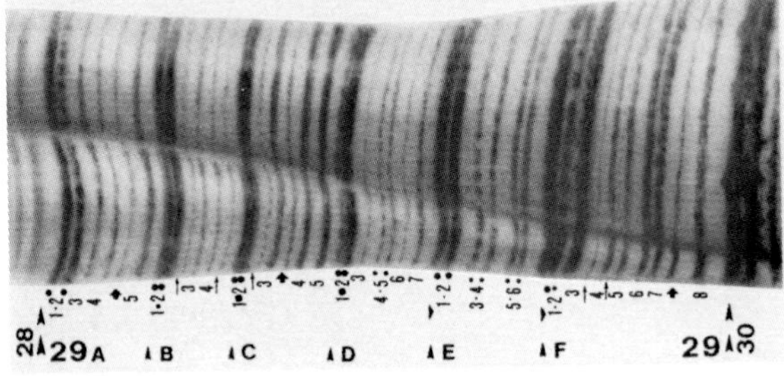

FIGURE 1 (continued)

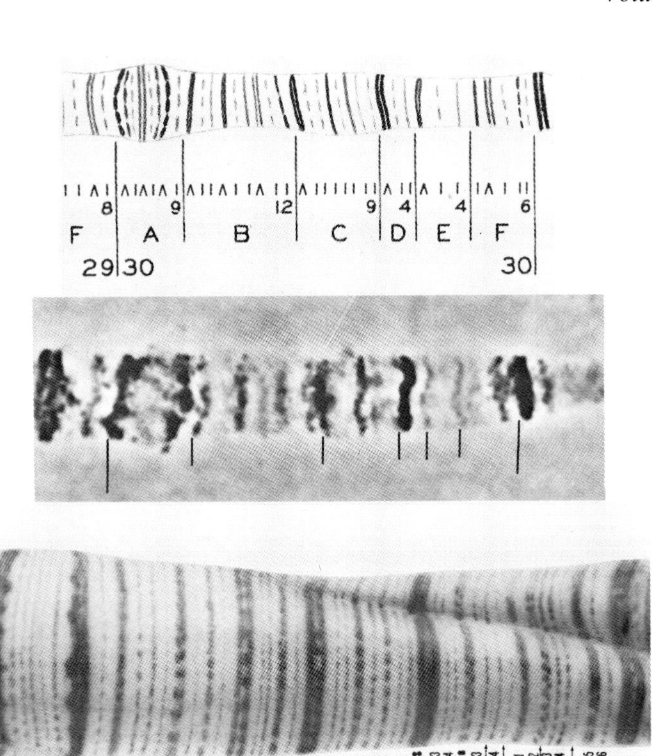

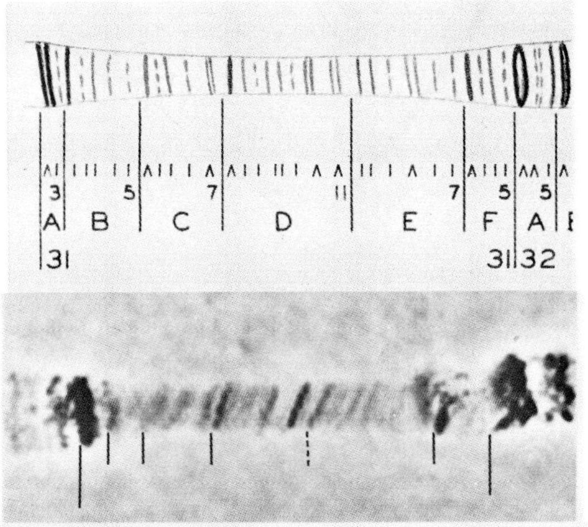

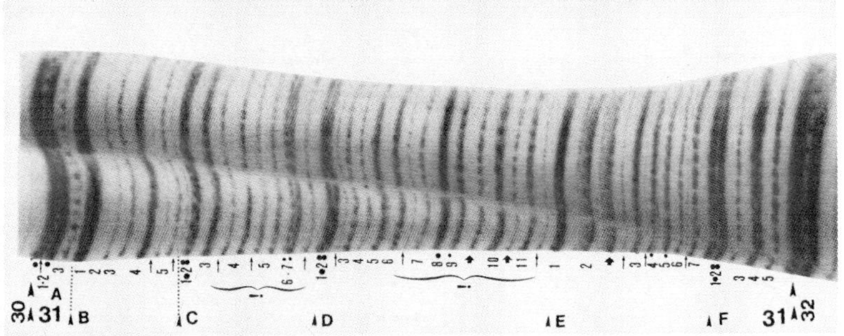

FIGURE 1 (continued)

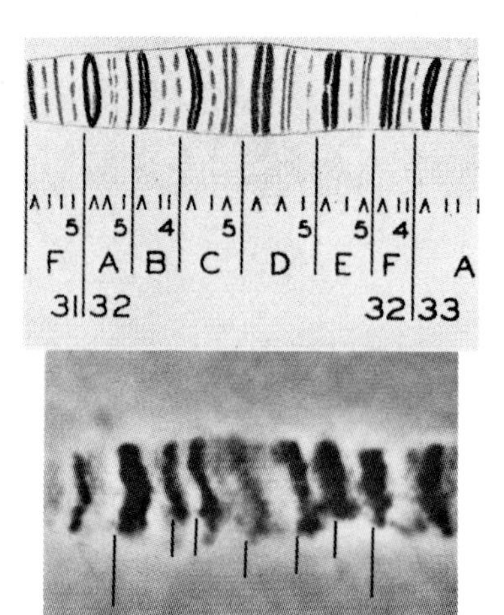

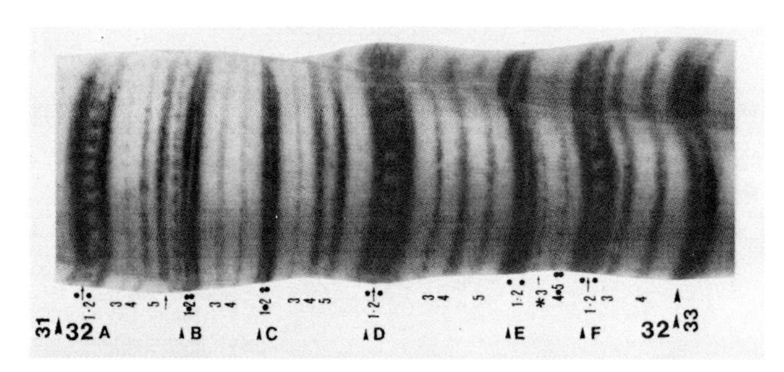

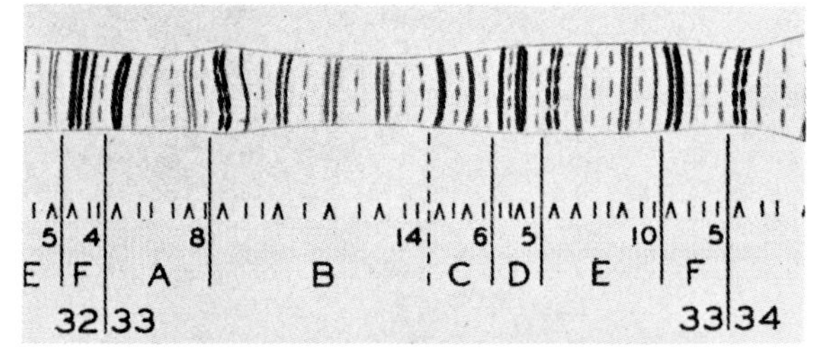

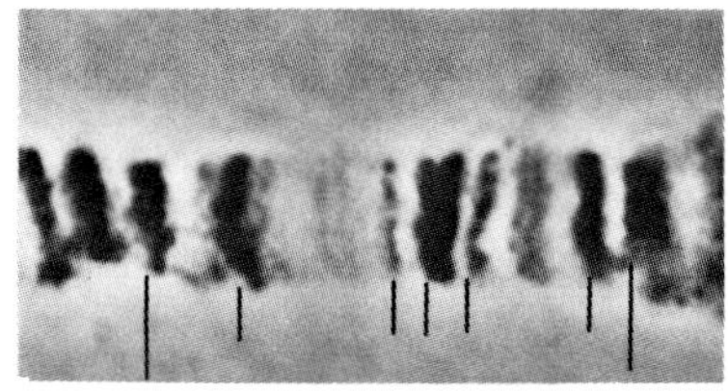

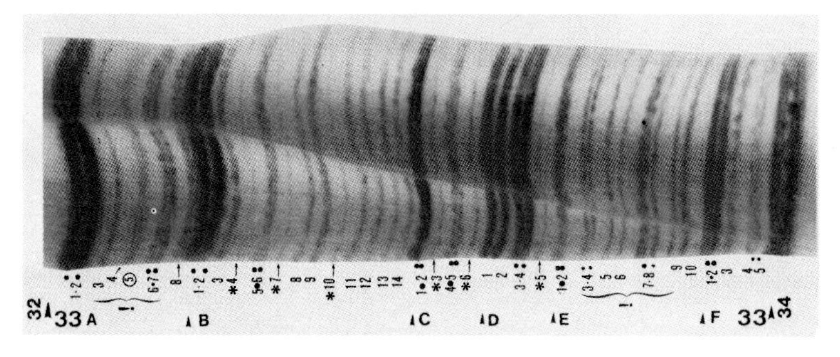

FIGURE 1 (continued)

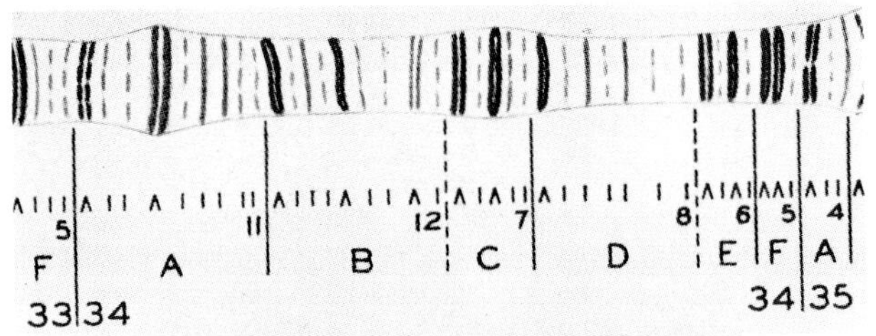

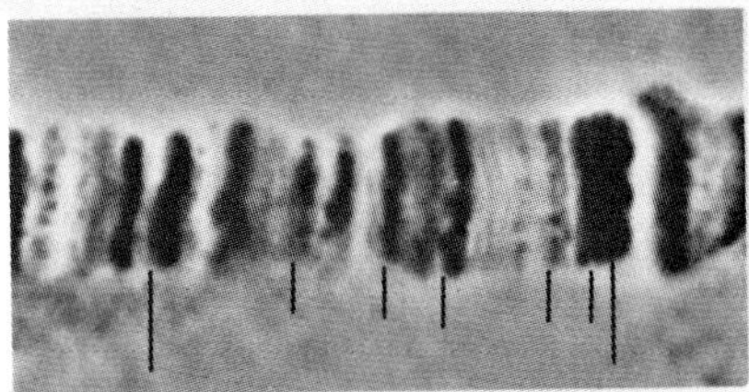

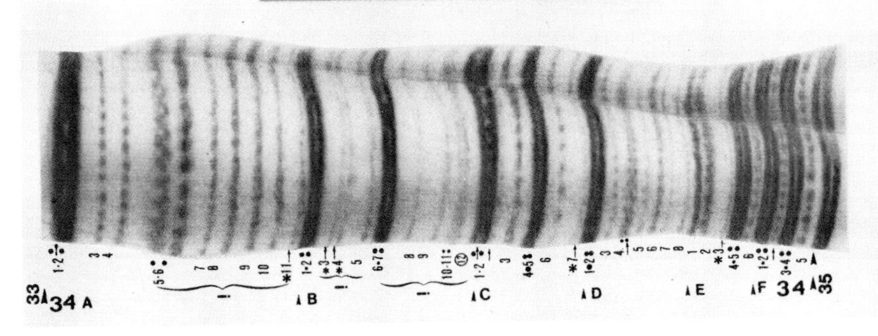

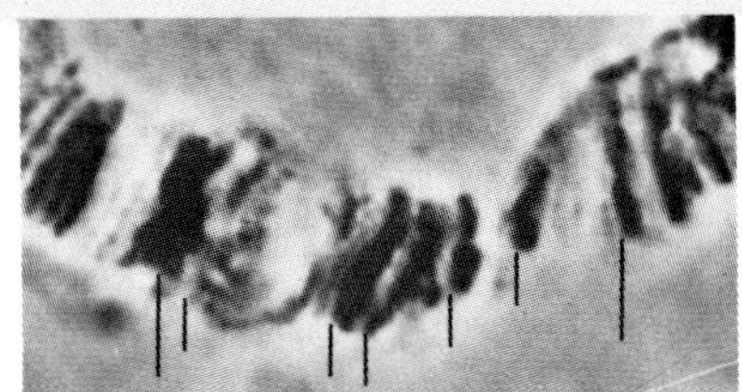

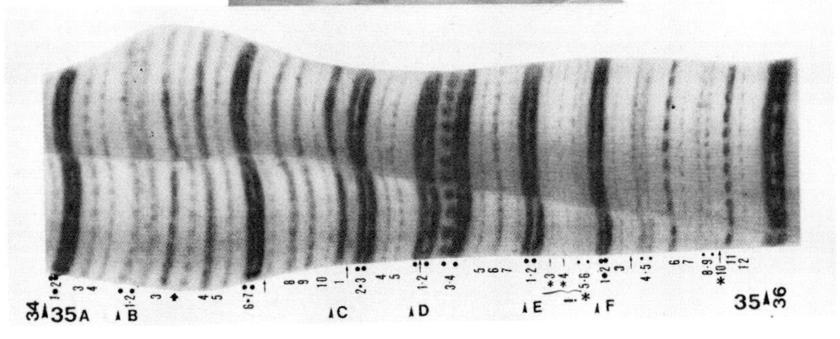

FIGURE 1 (continued)

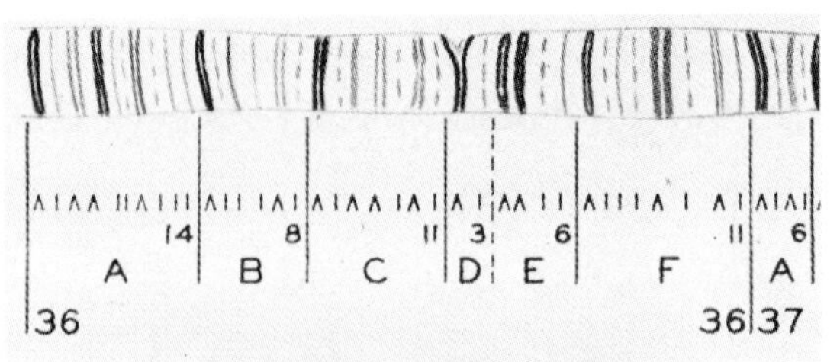

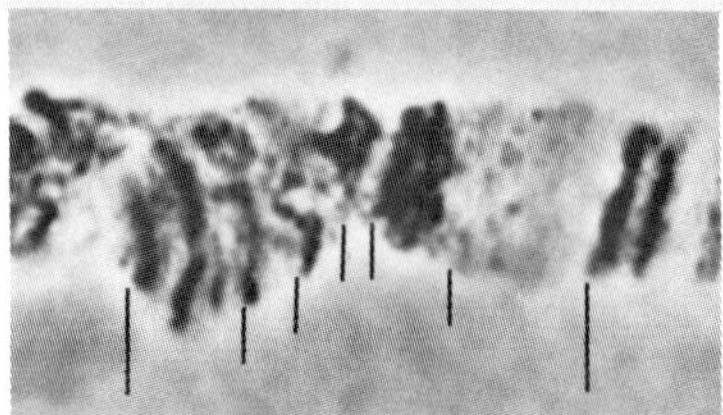

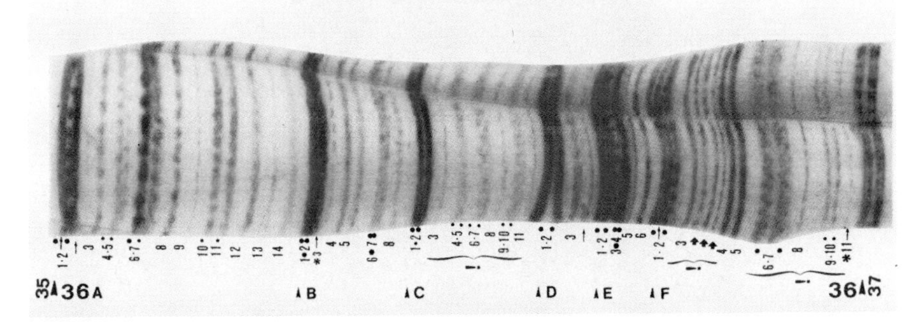

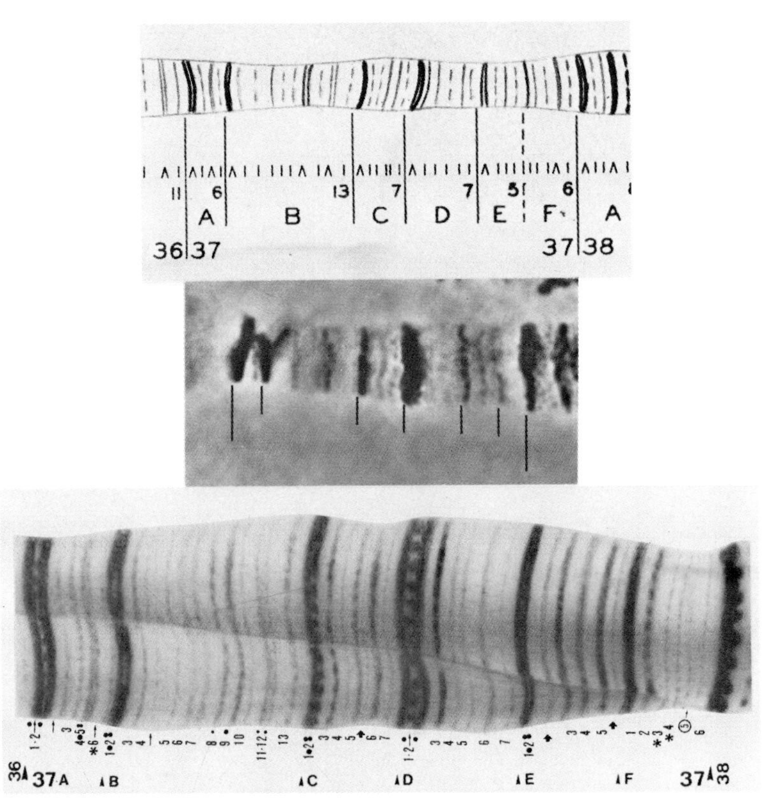

FIGURE 1 (continued)

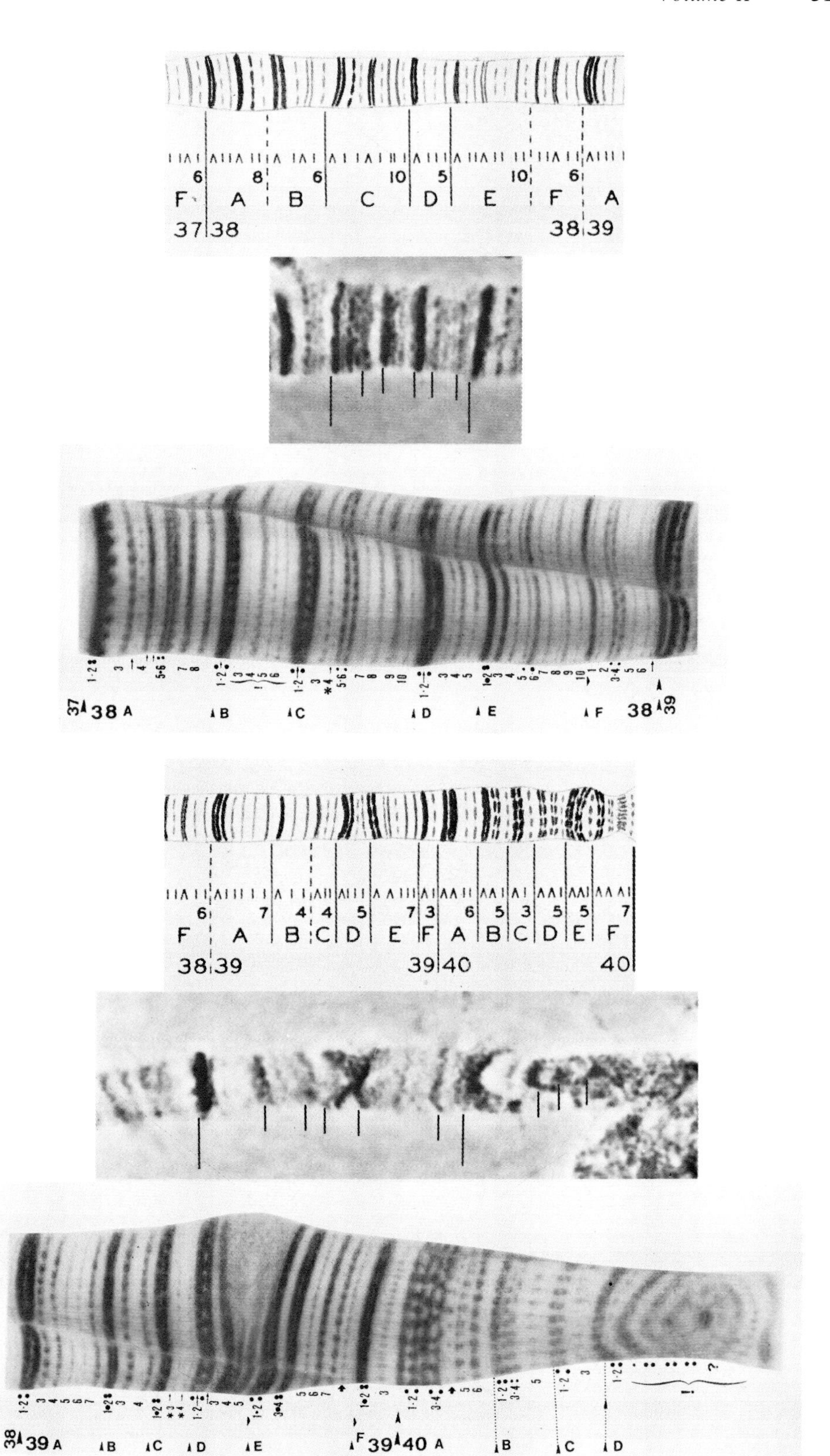

FIGURE 1 (continued)

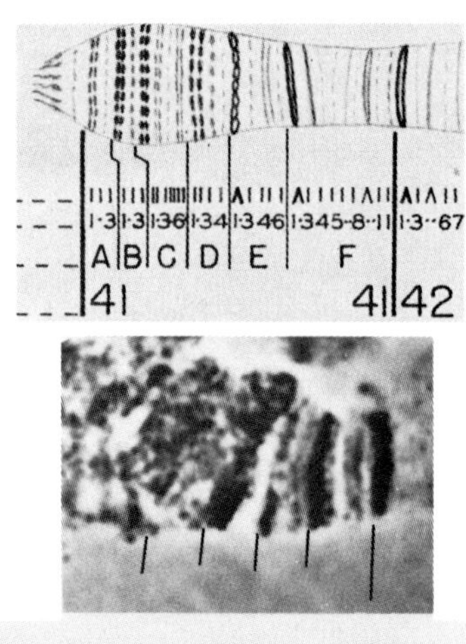

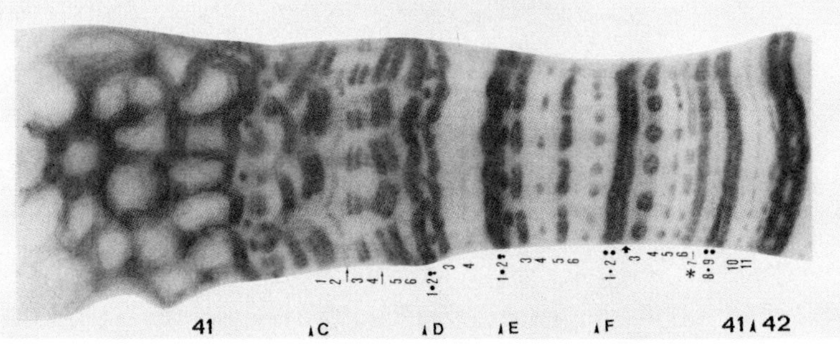

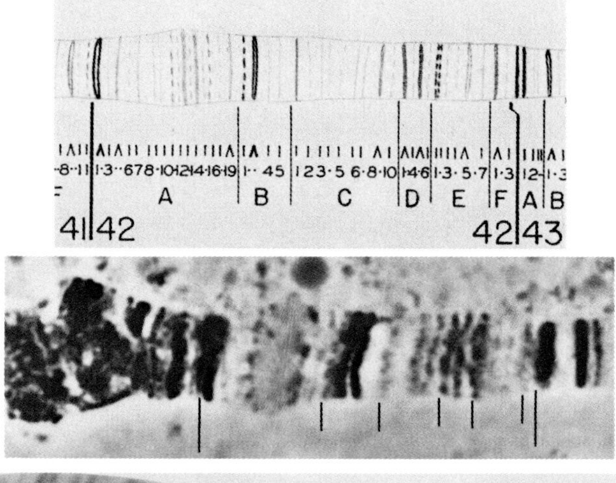

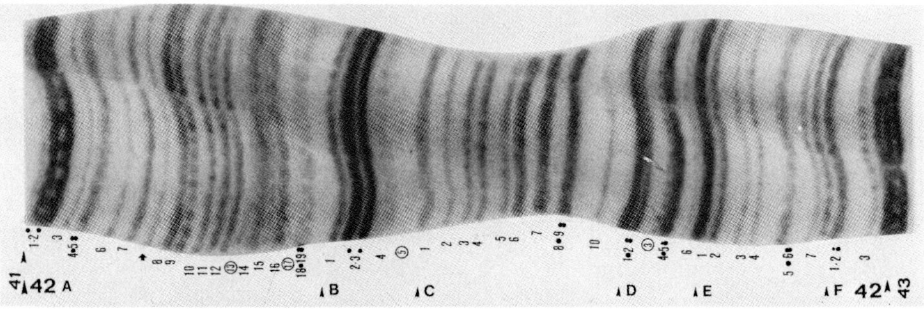

FIGURE 1 (continued)

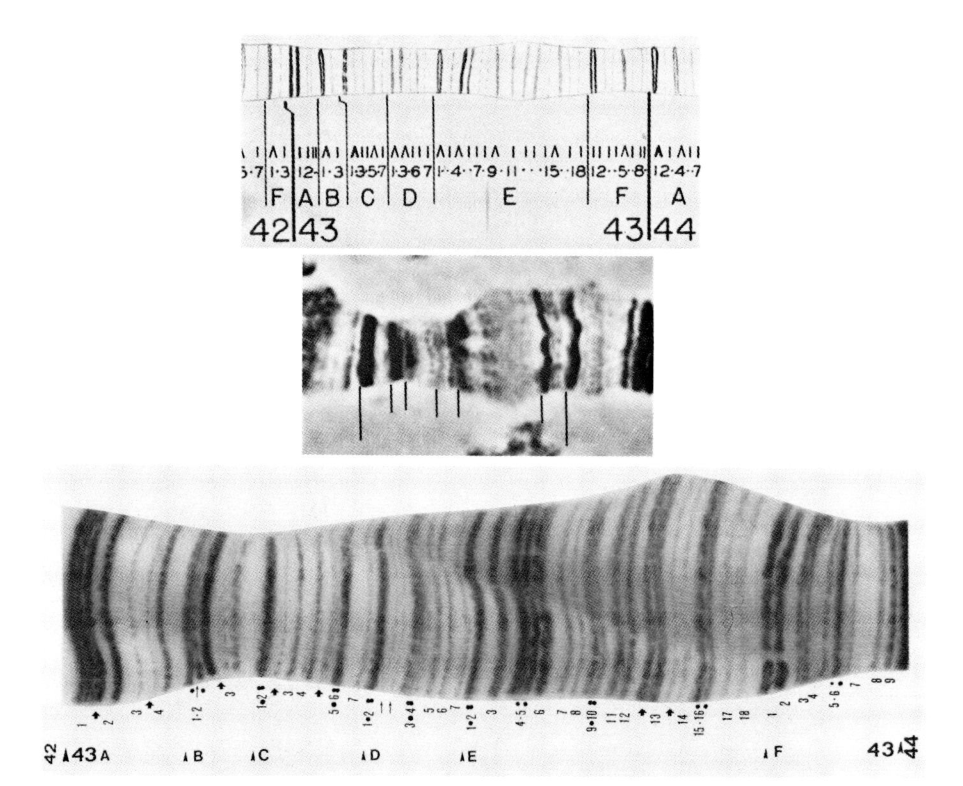

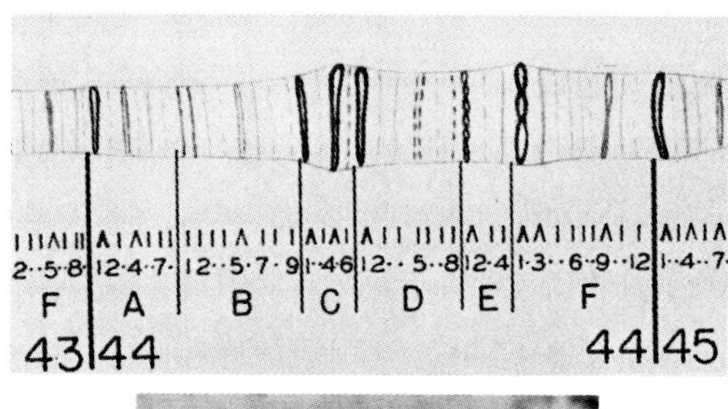

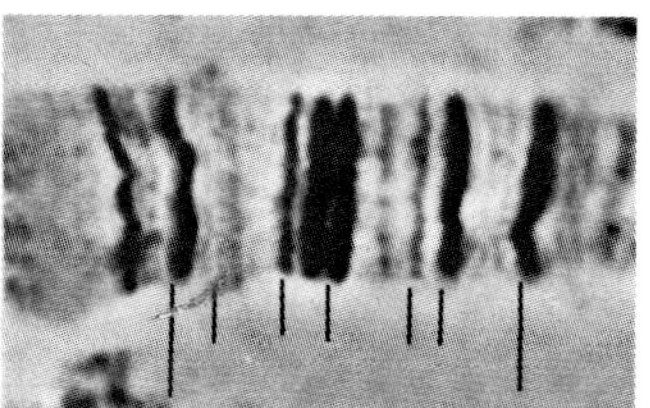

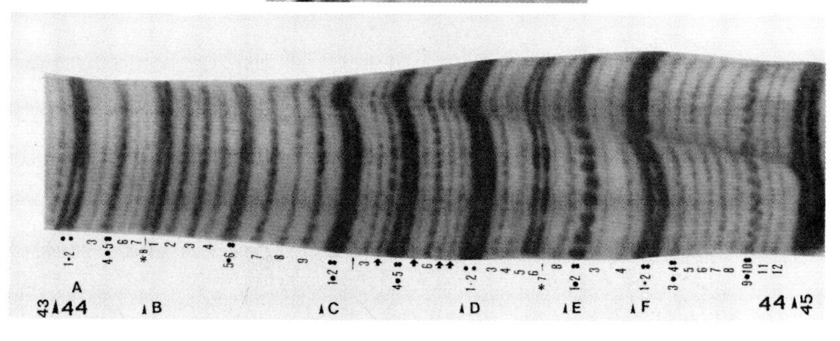

FIGURE 1 (continued)

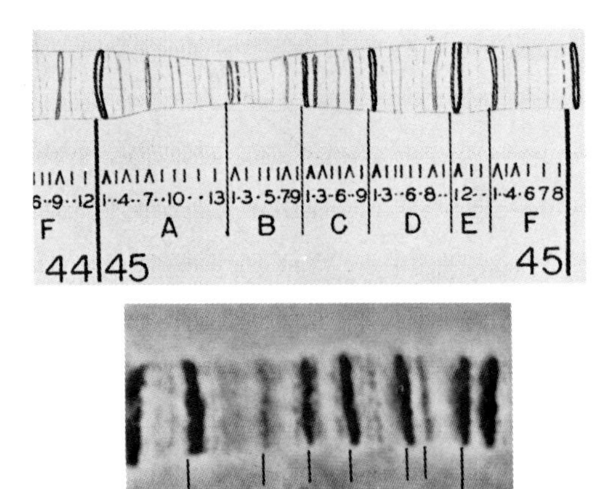

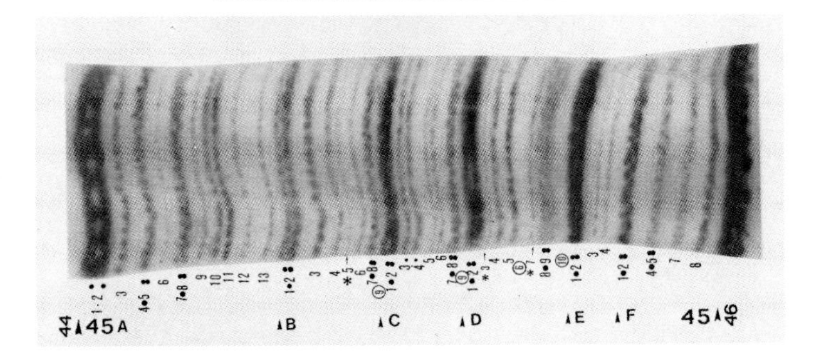

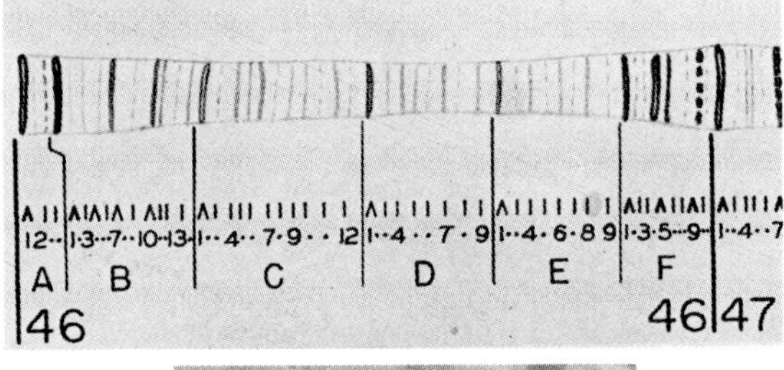

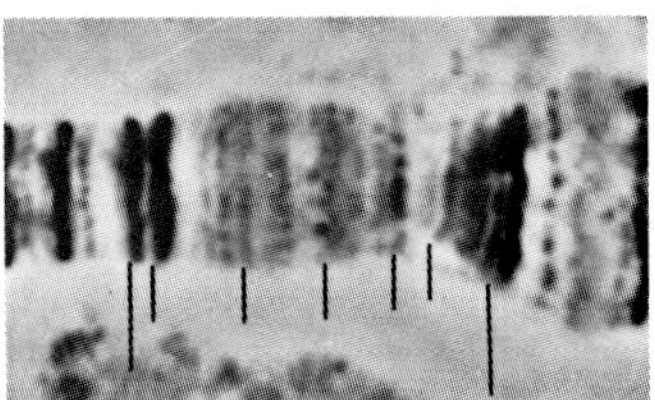

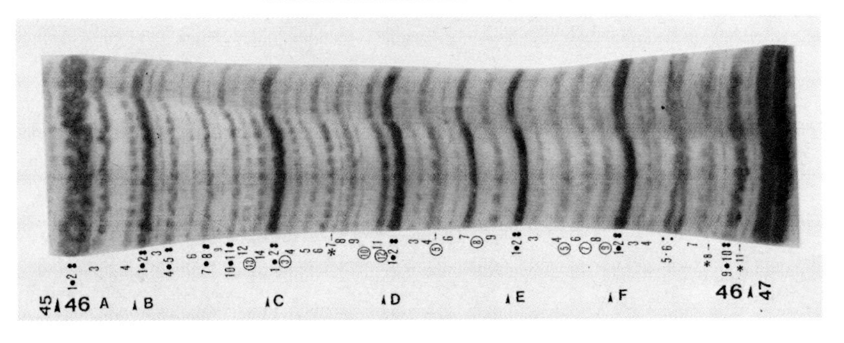

FIGURE 1 (continued)

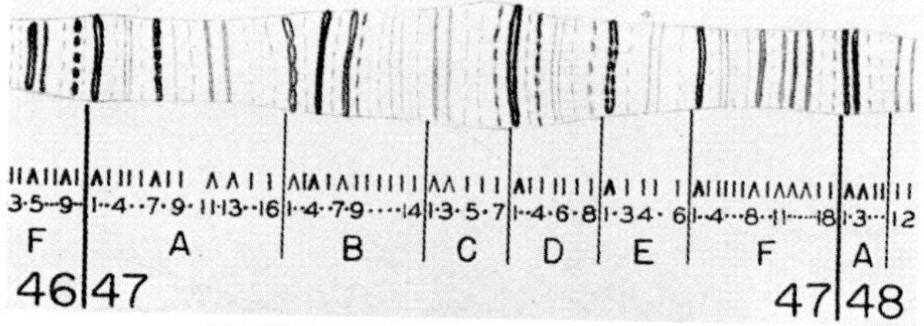

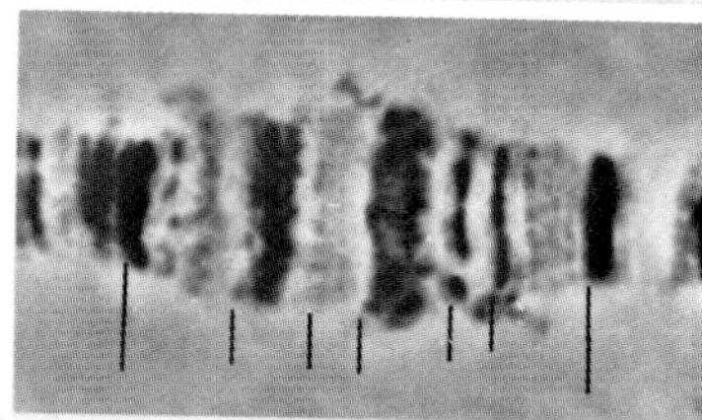

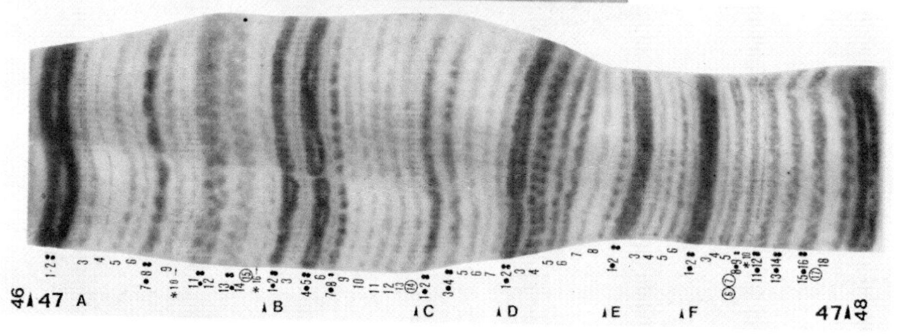

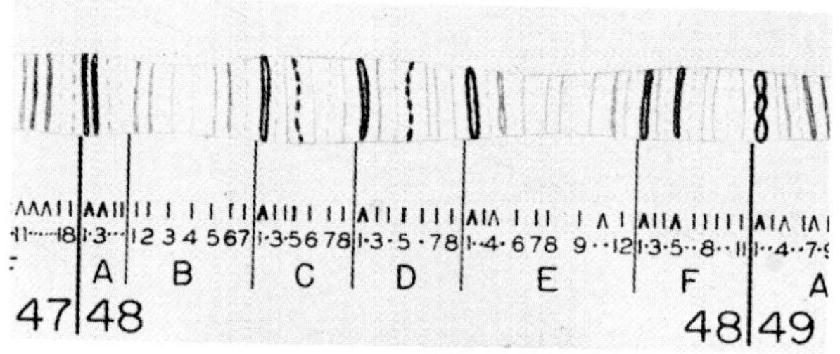

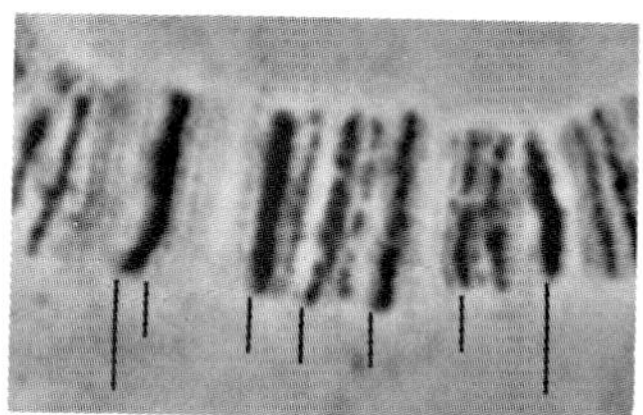

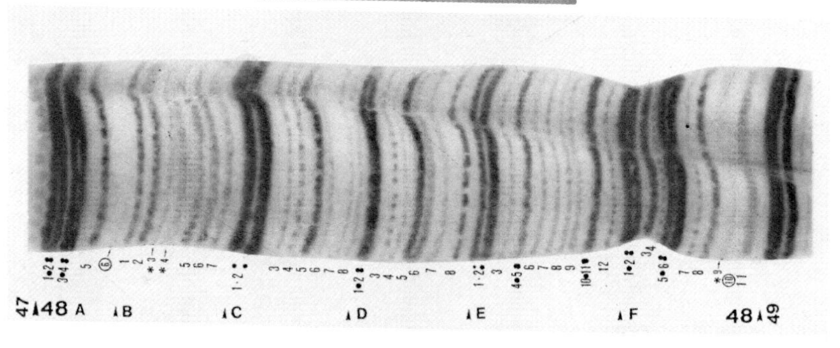

FIGURE 1 (continued)

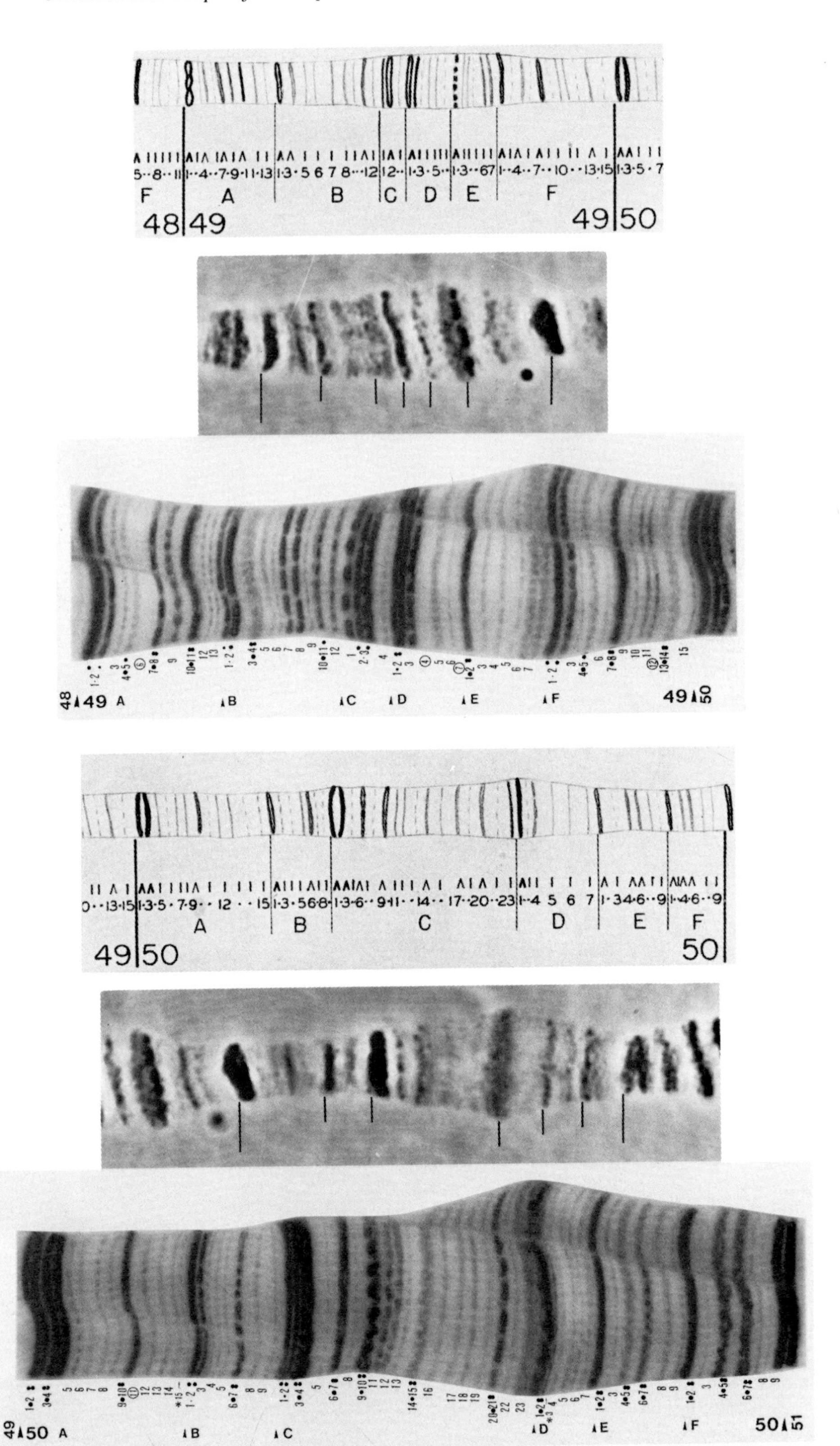

FIGURE 1 (continued)

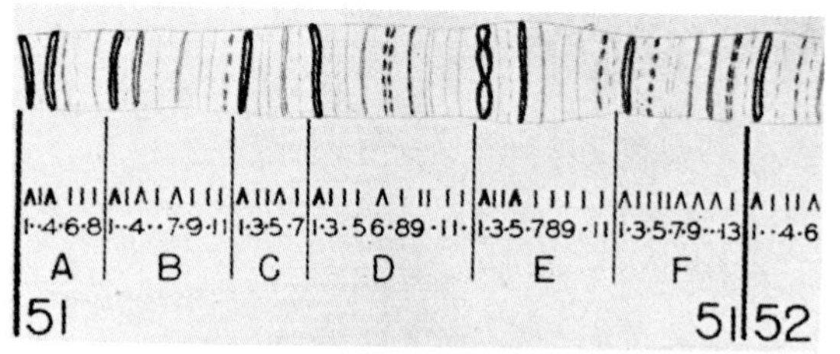

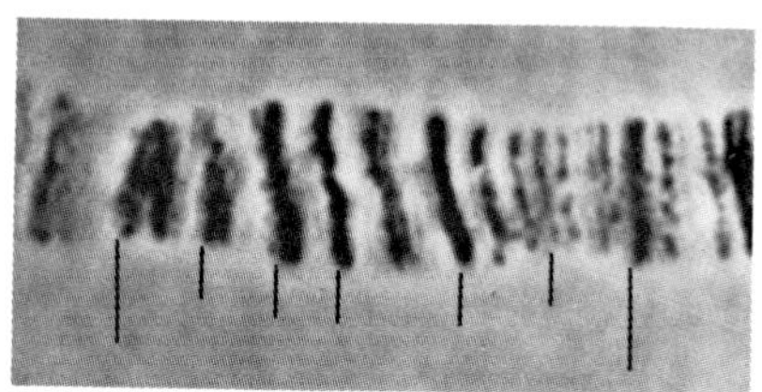

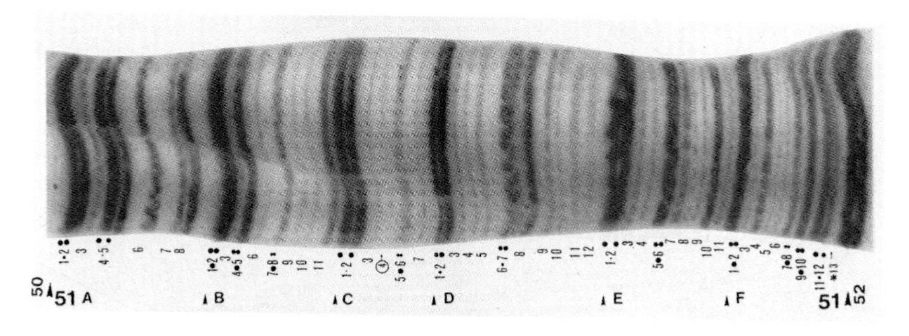

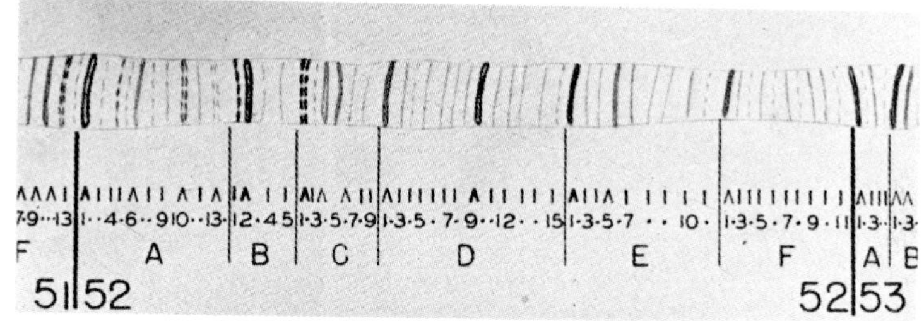

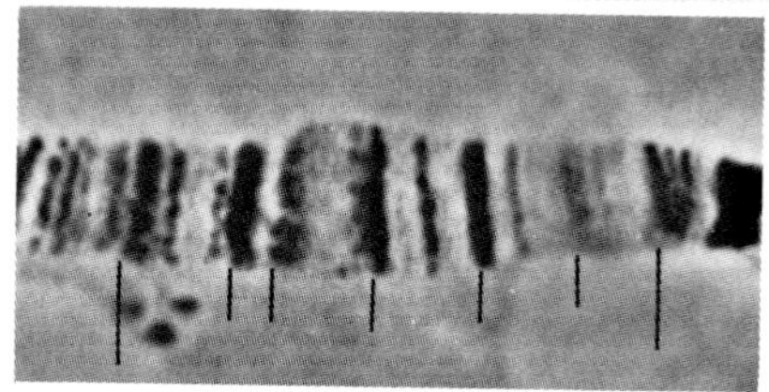

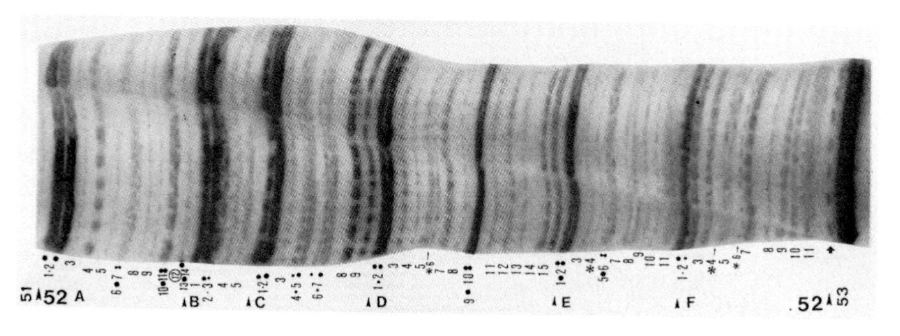

FIGURE 1 (continued)

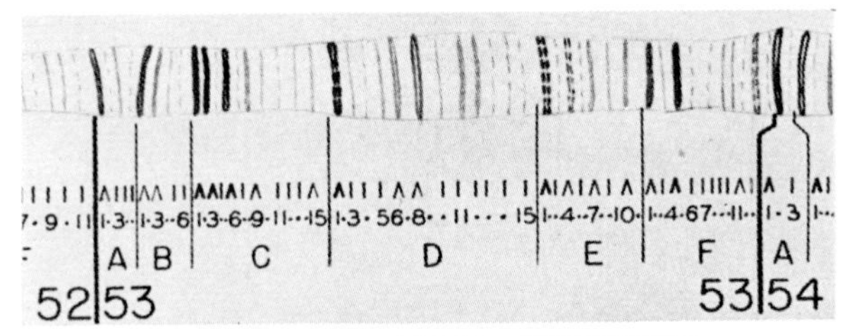

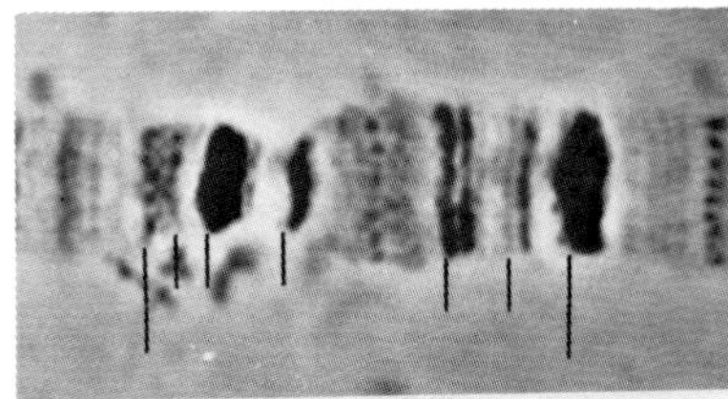

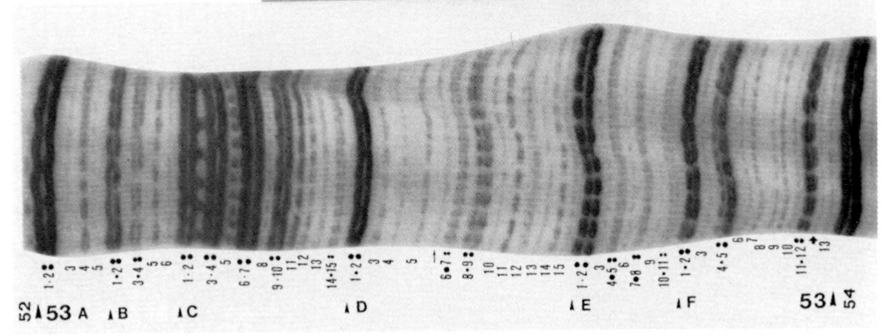

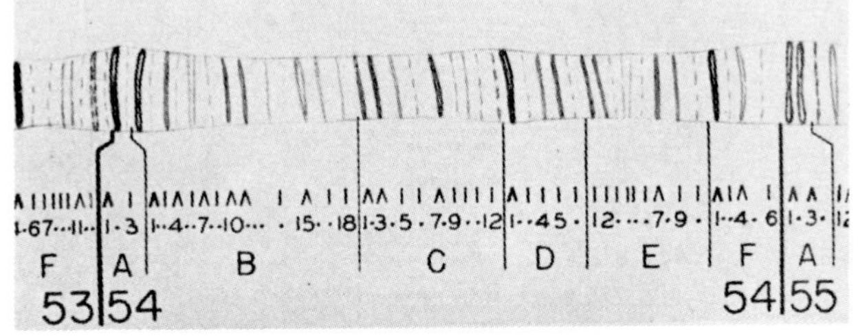

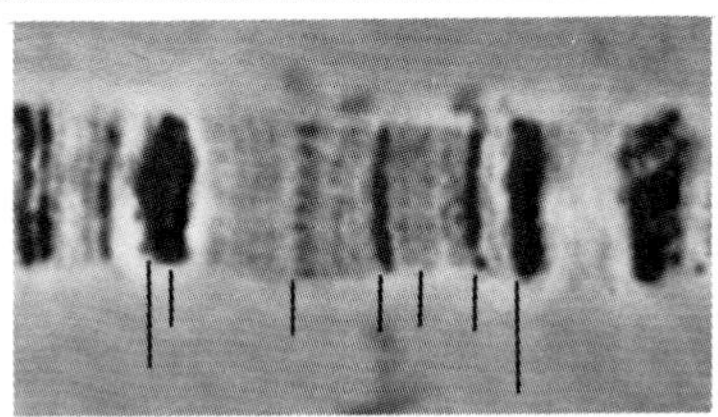

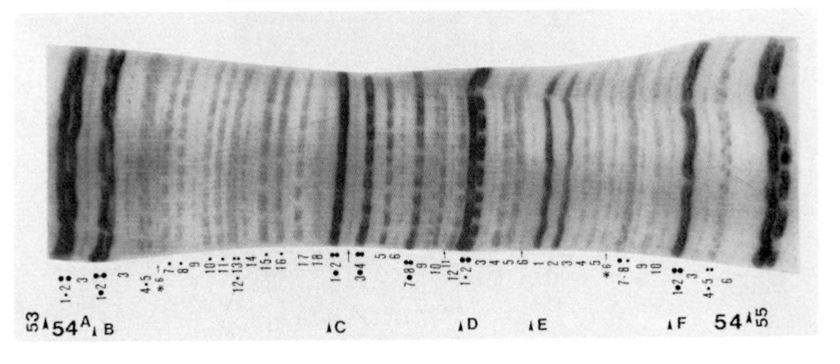

FIGURE 1 (continued)

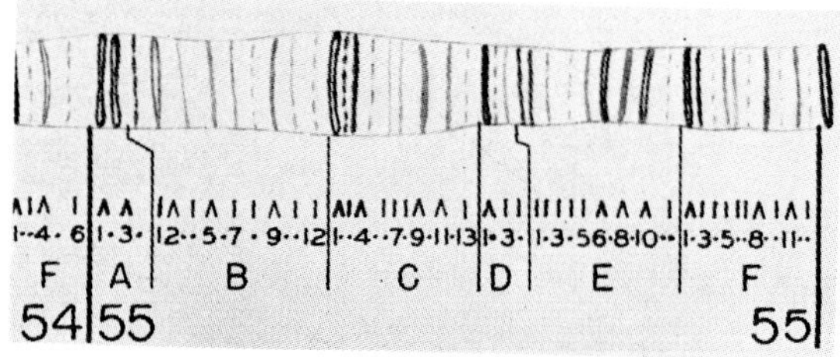

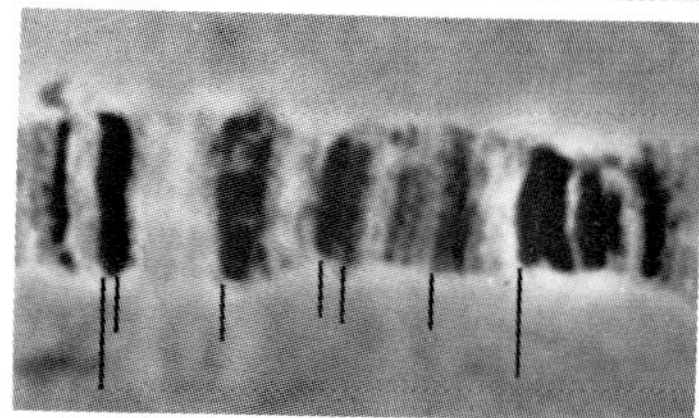

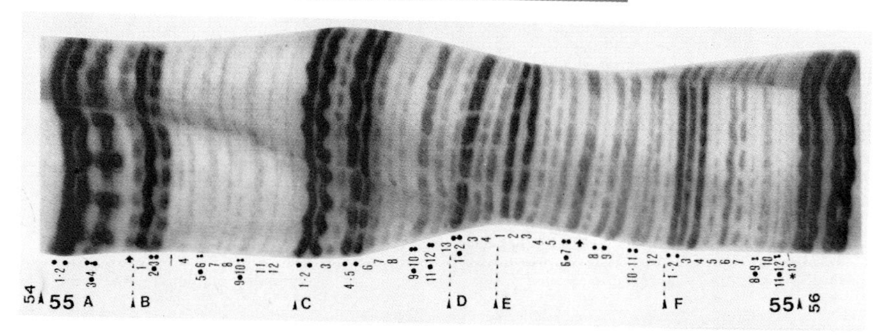

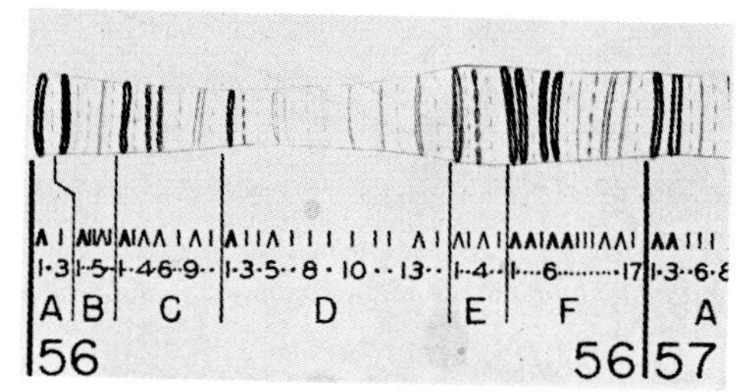

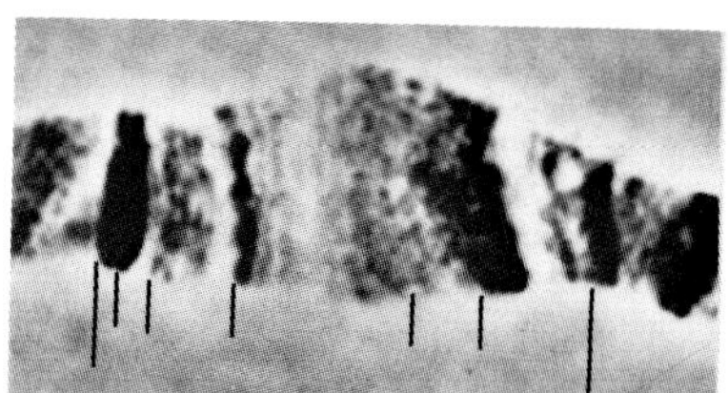

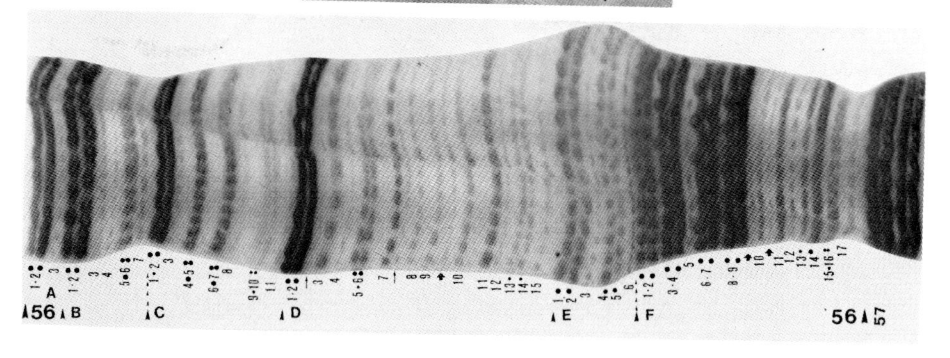

FIGURE 1 (continued)

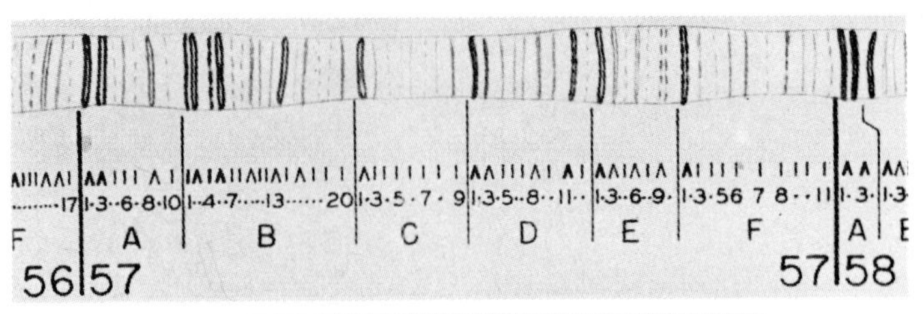

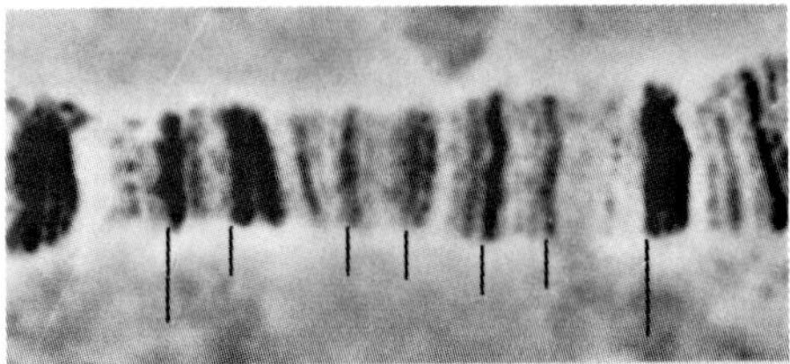

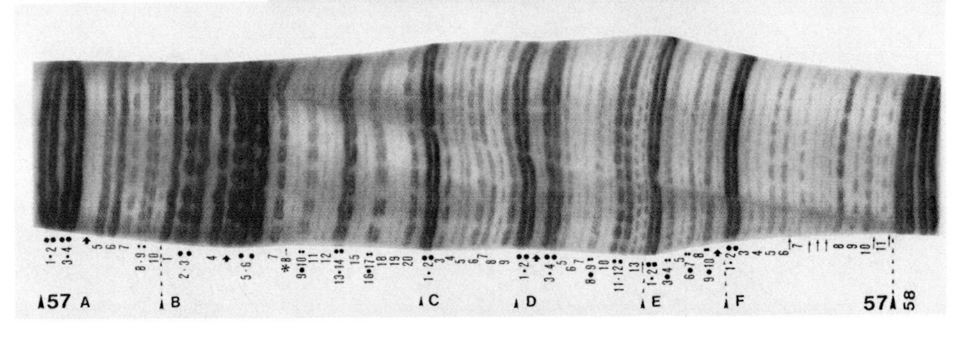

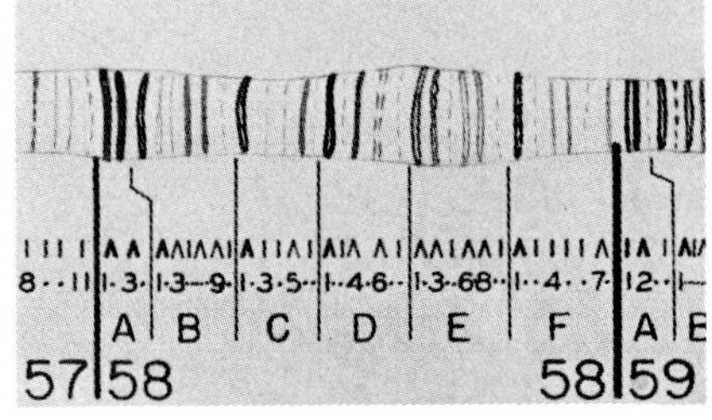

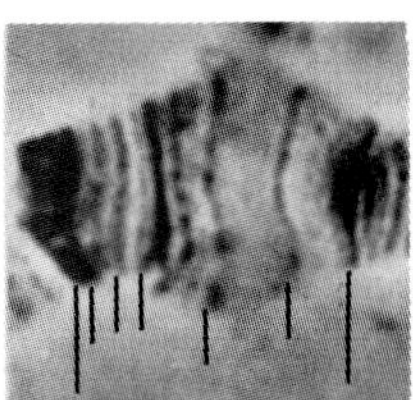

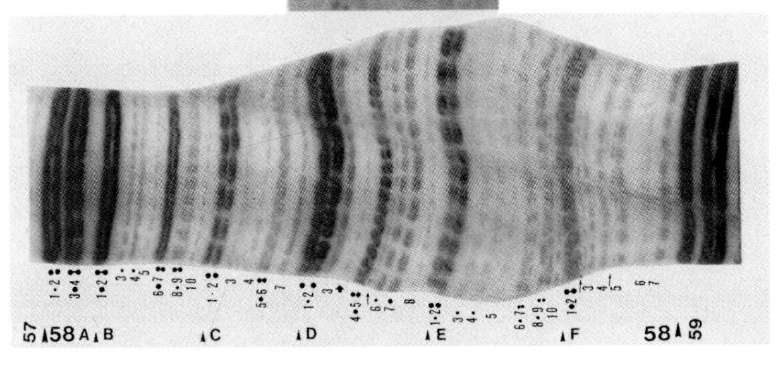

FIGURE 1 (continued)

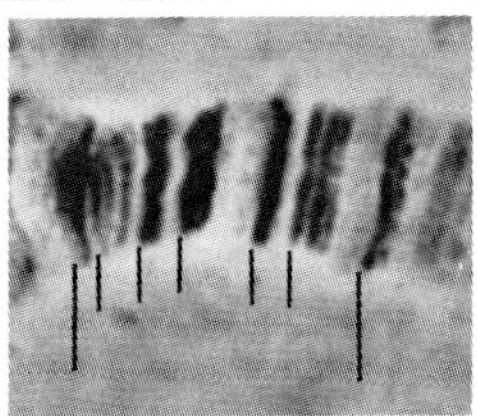

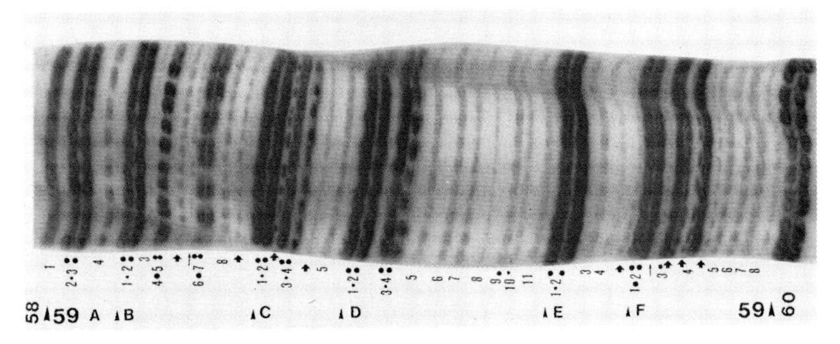

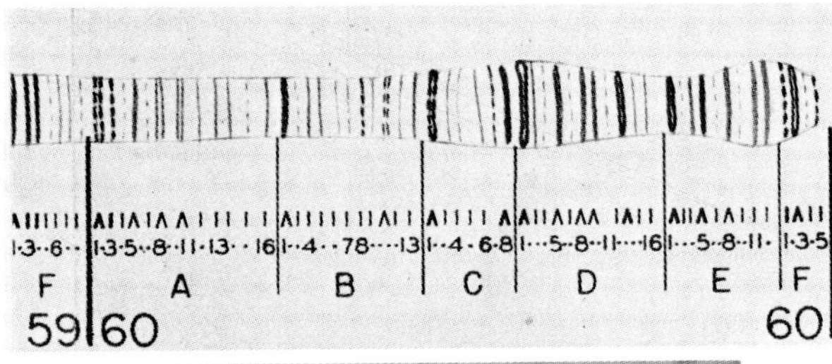

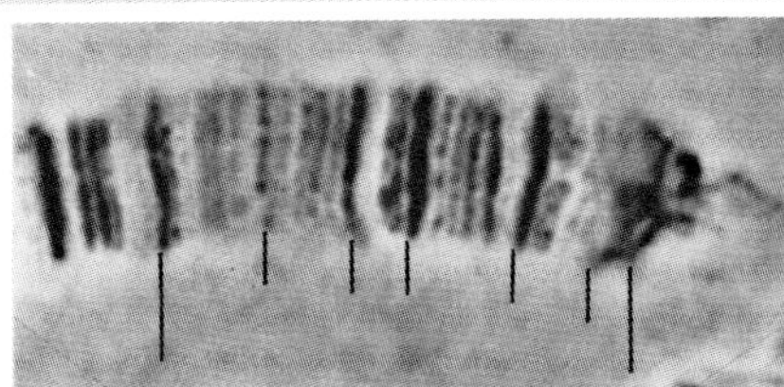

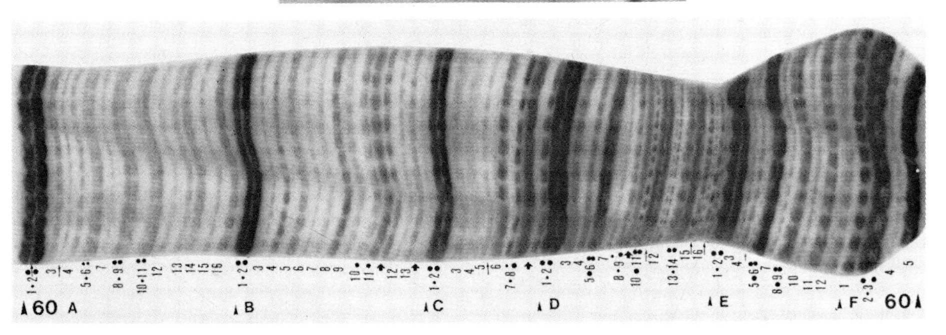

FIGURE 1 (continued)

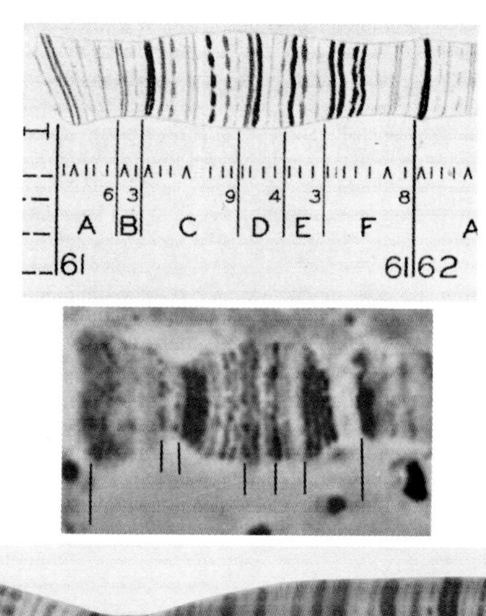

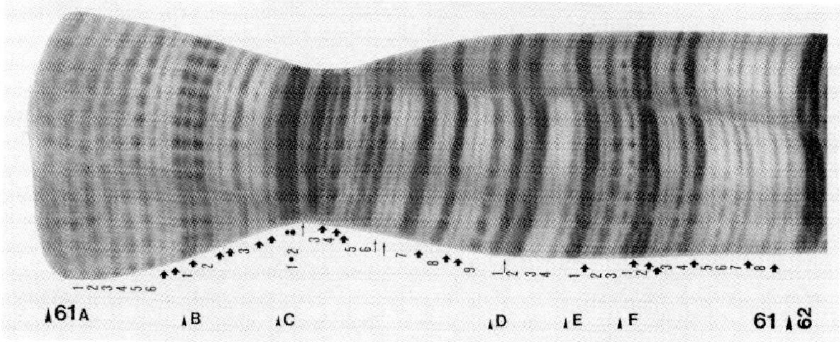

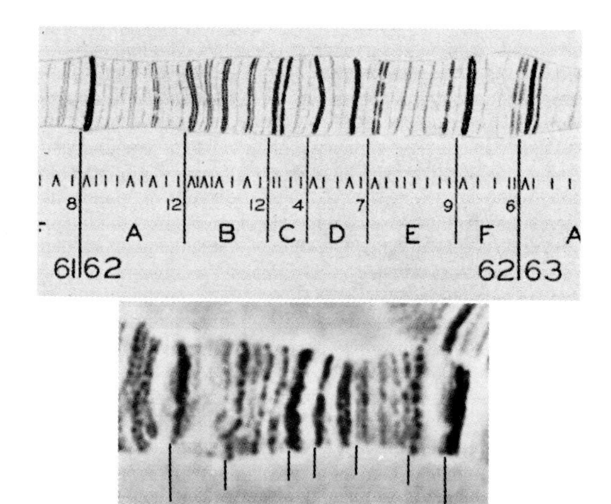

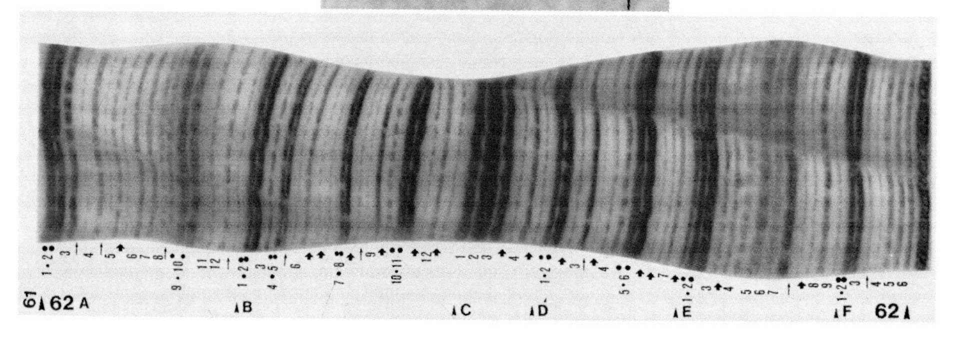

FIGURE 1 (continued)

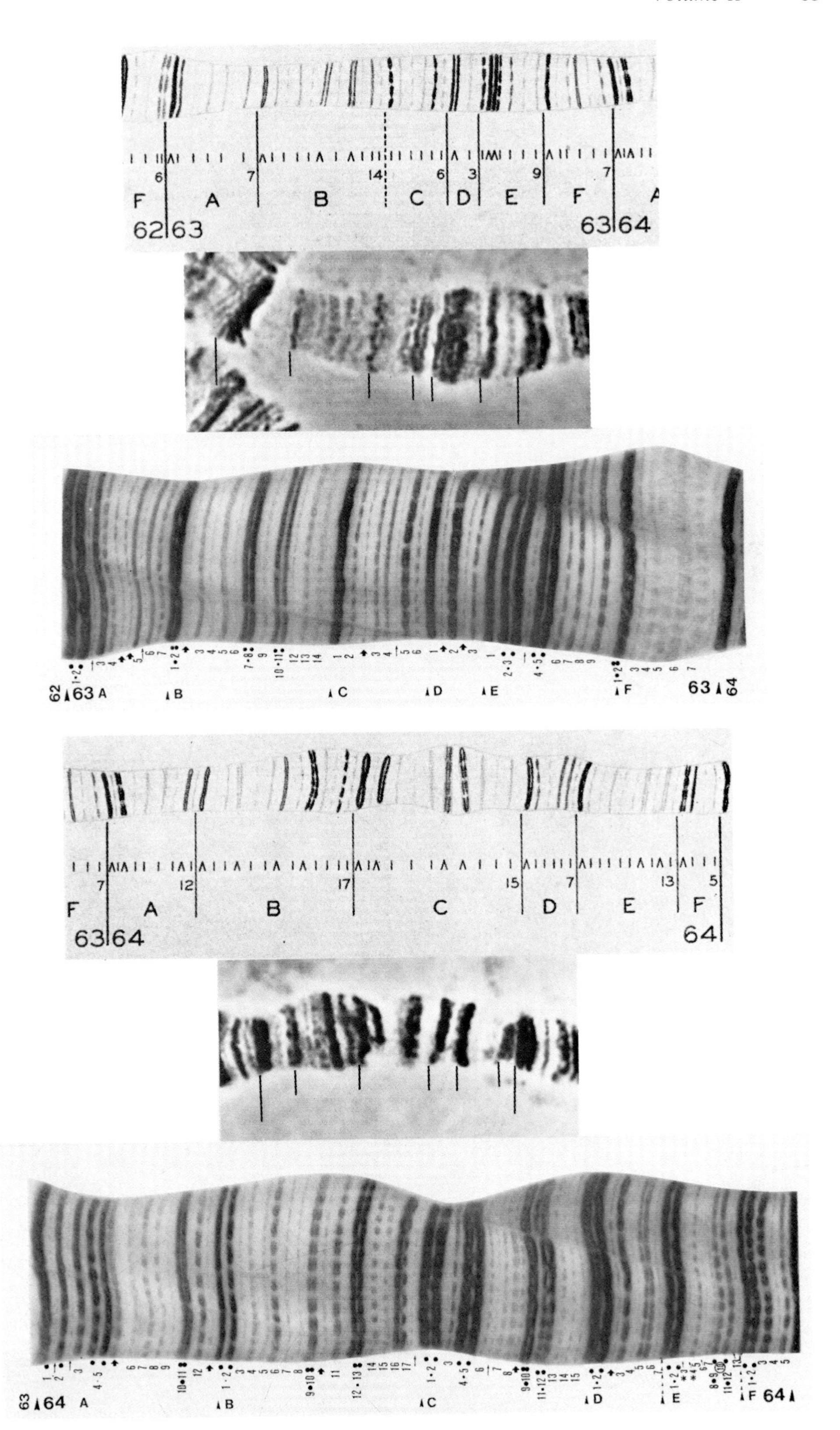

FIGURE 1 (continued)

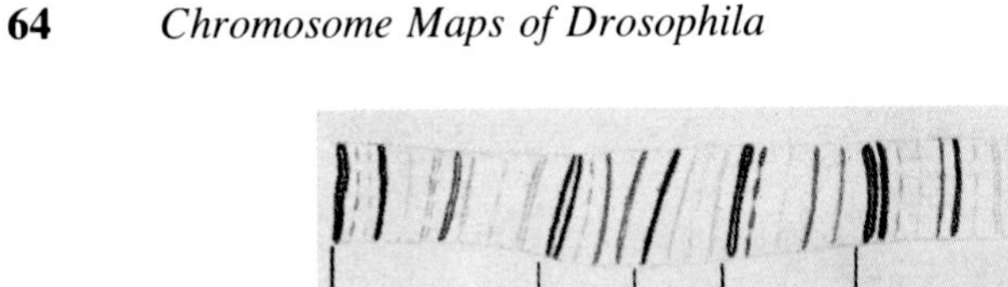

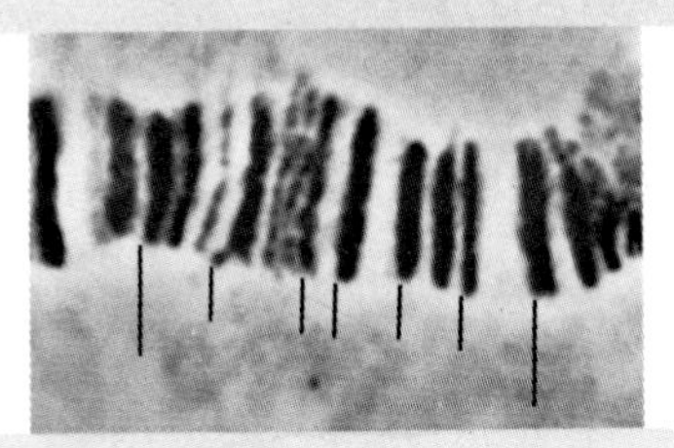

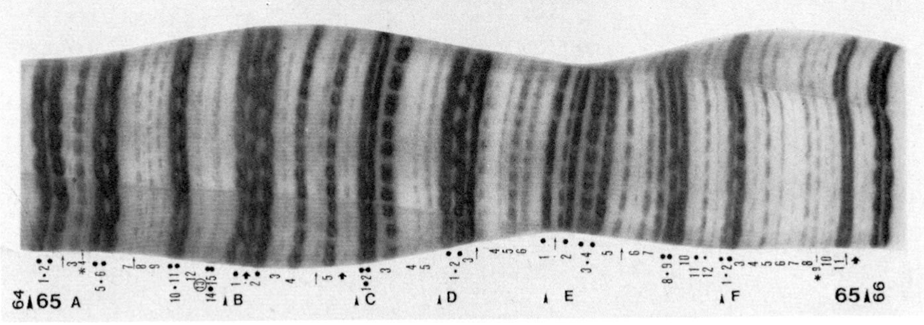

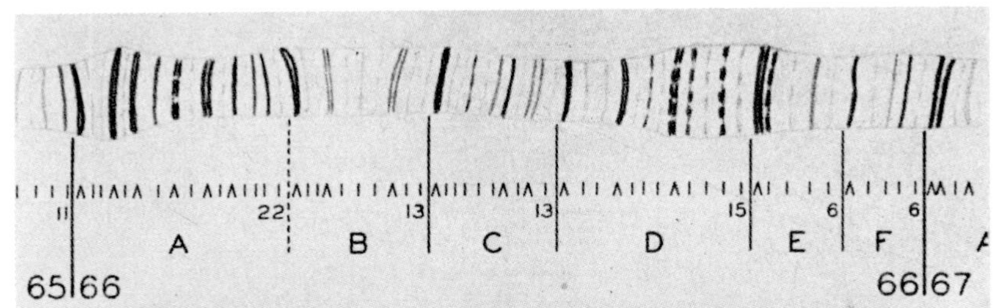

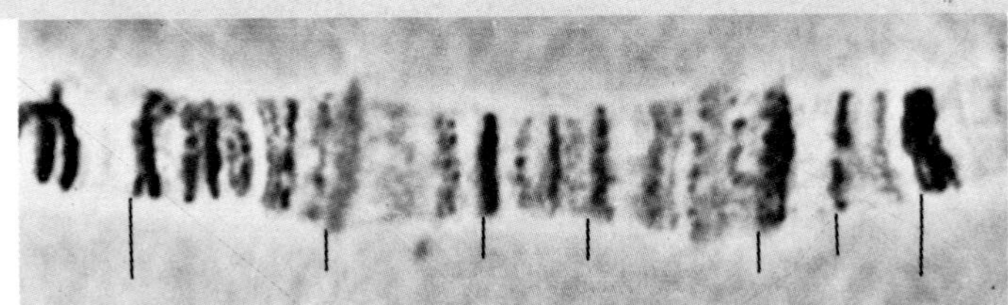

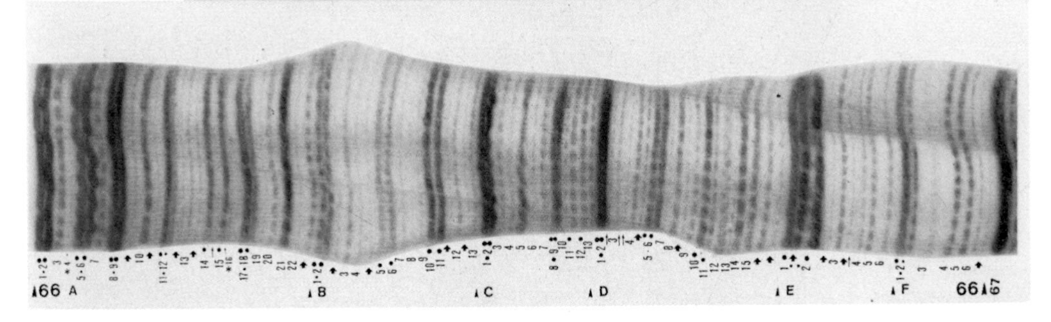

FIGURE 1 (continued)

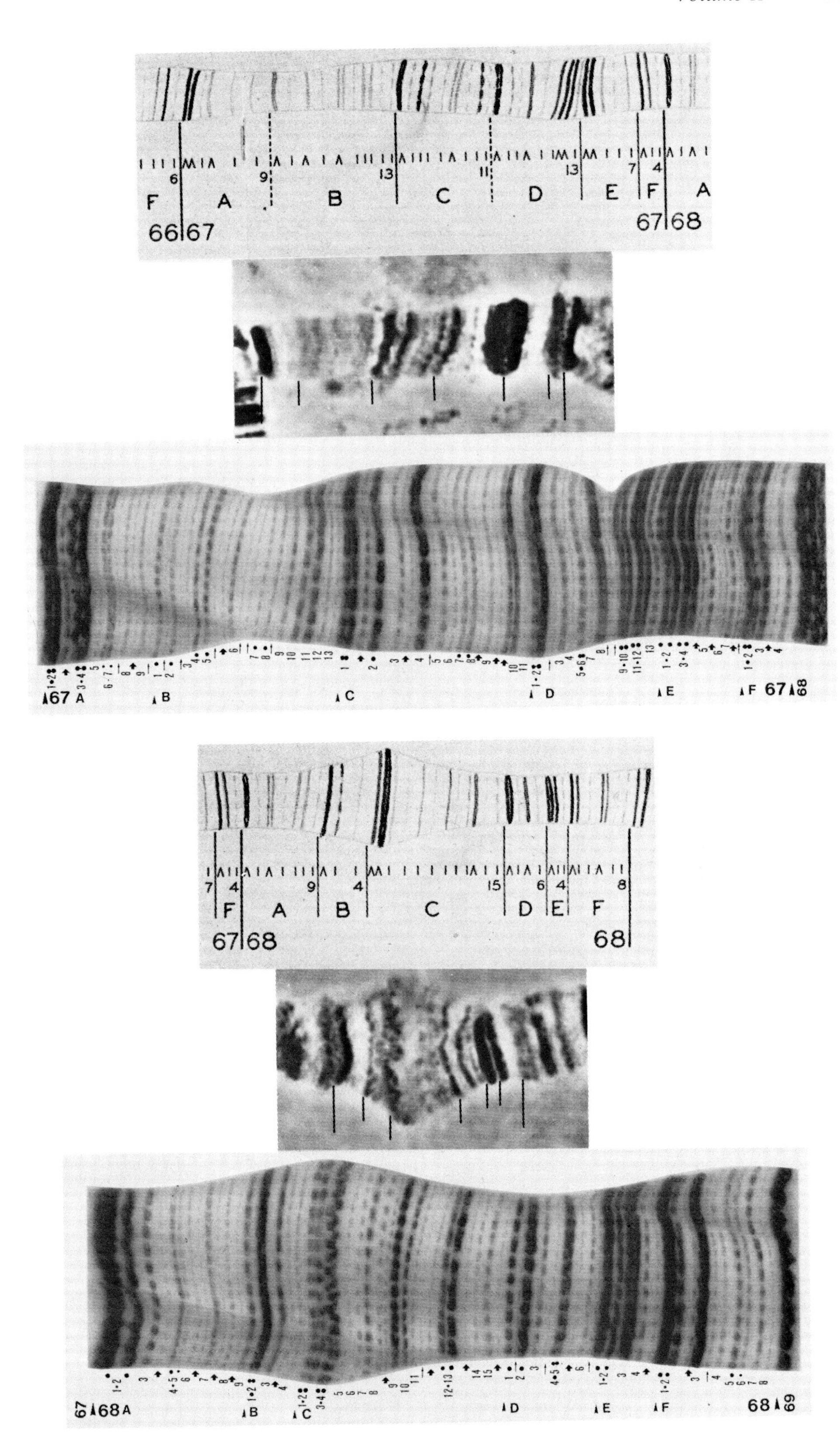

FIGURE 1 (continued)

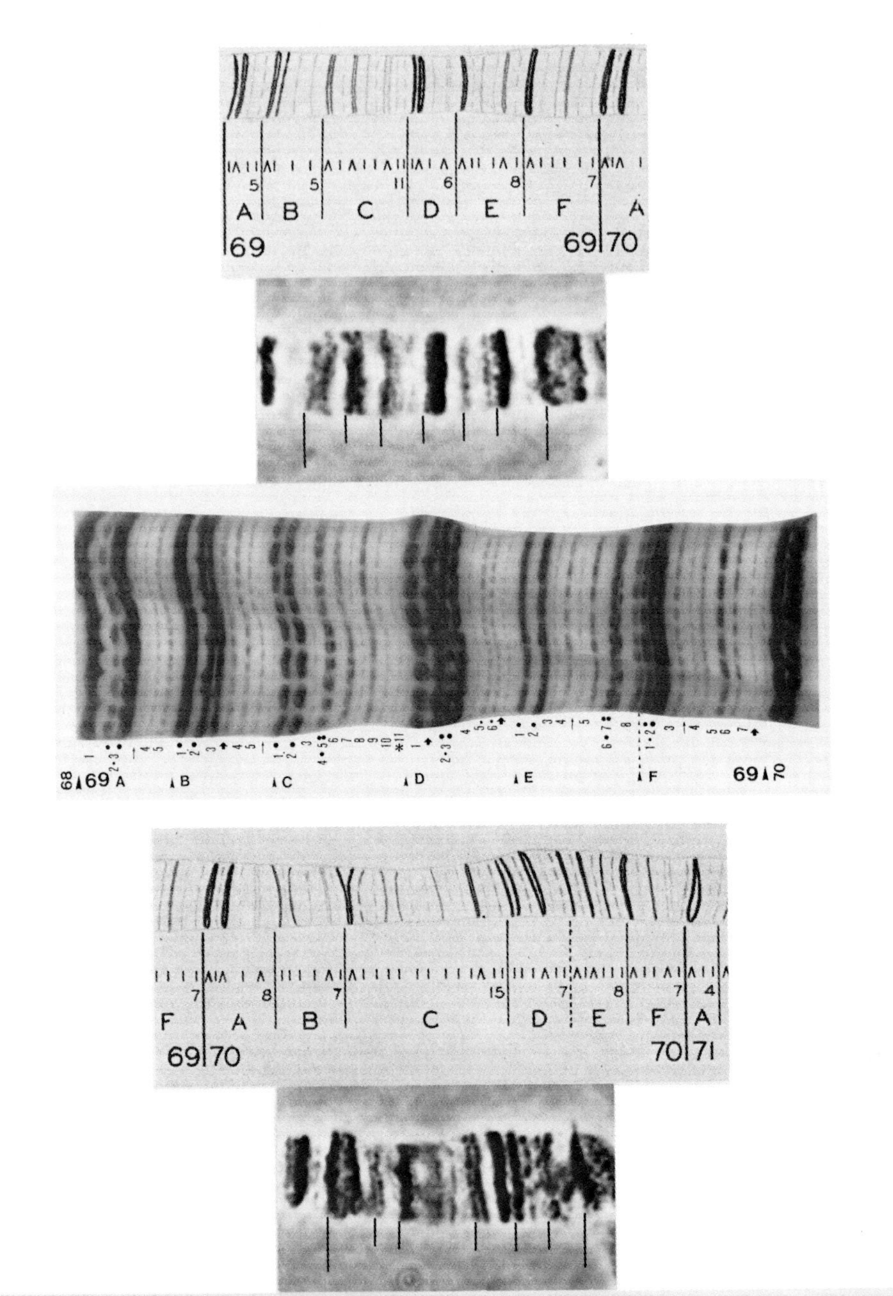

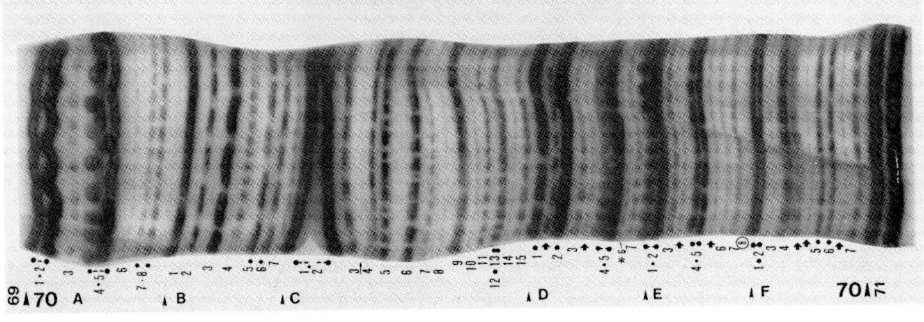

FIGURE 1 (continued)

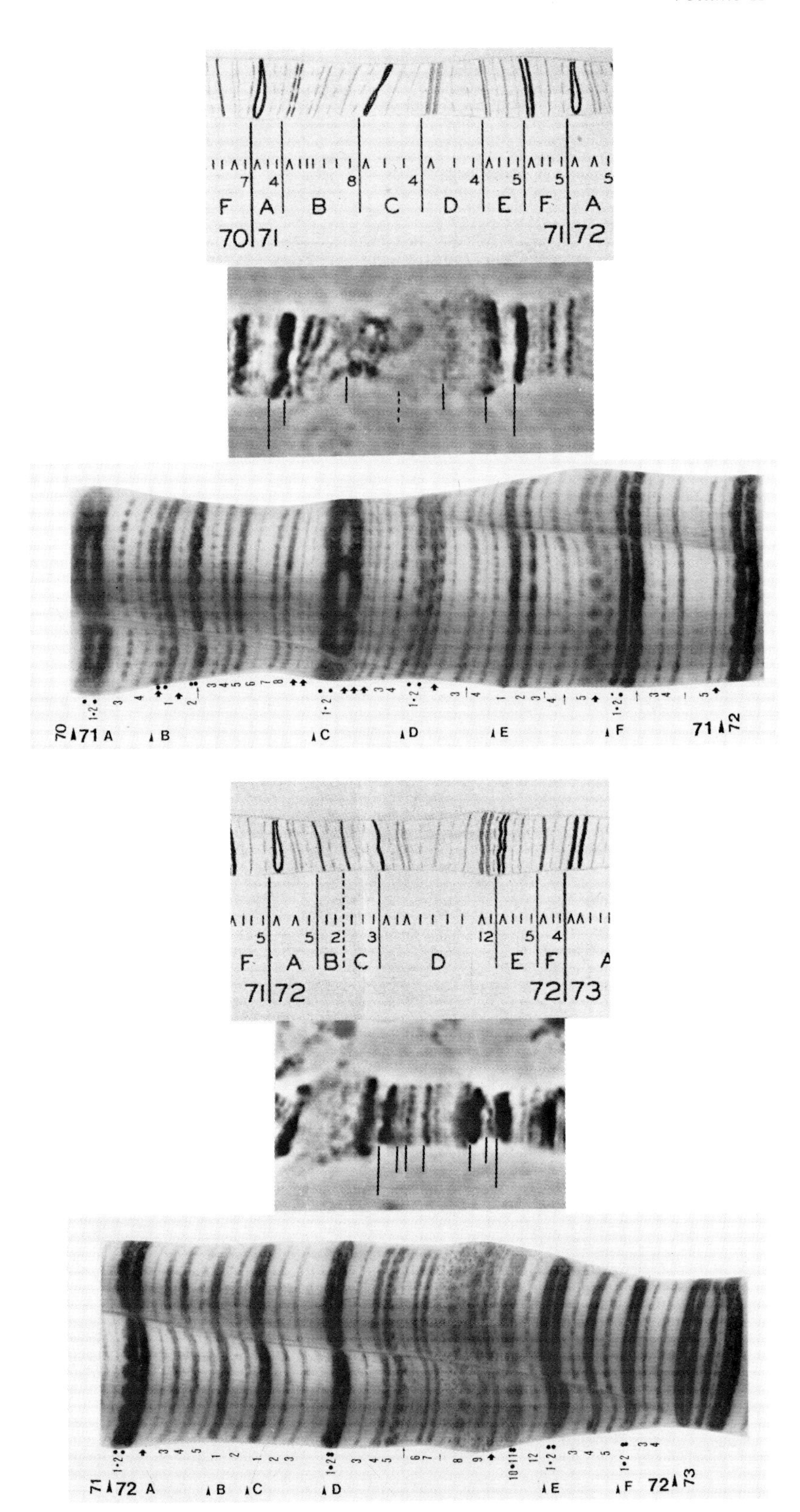

FIGURE 1 (continued)

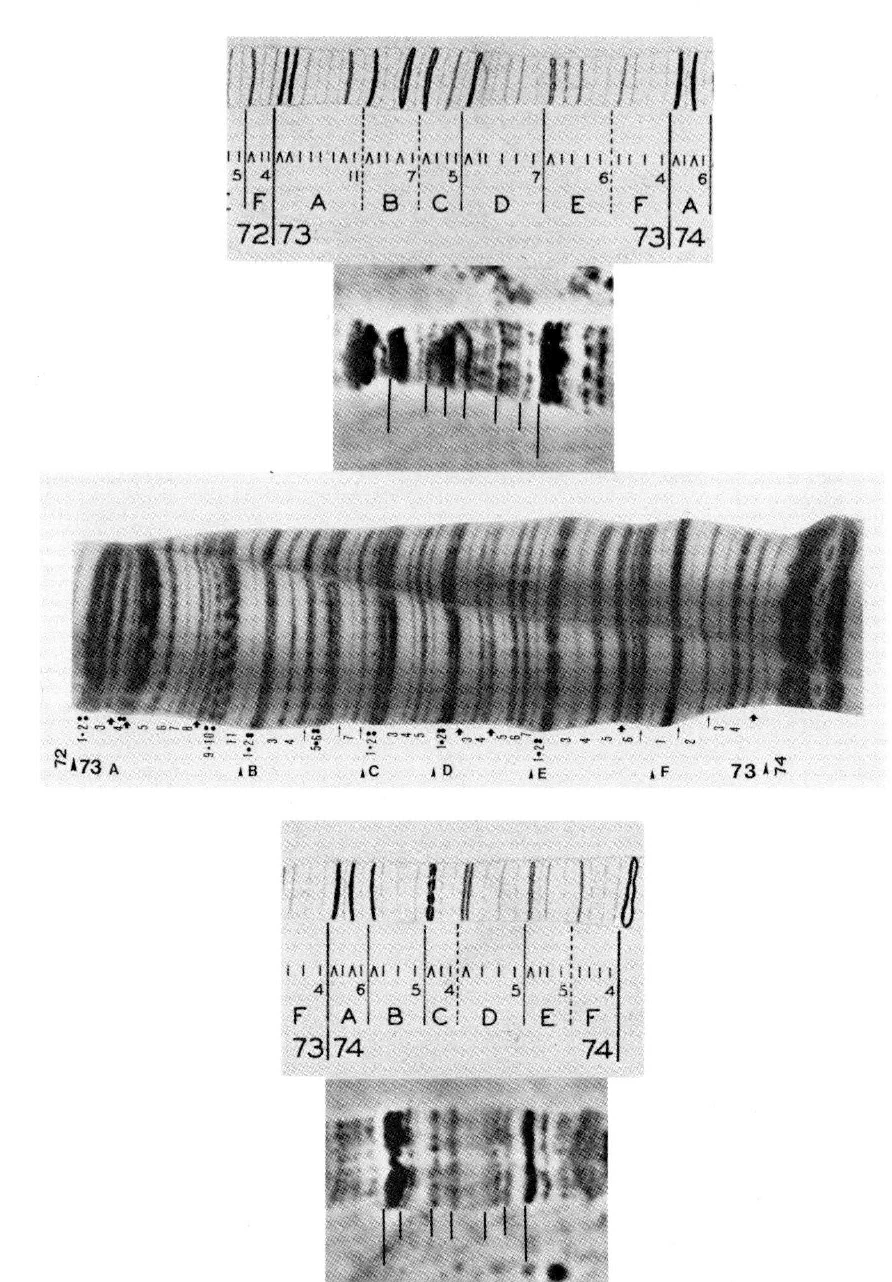

FIGURE 1 (continued)

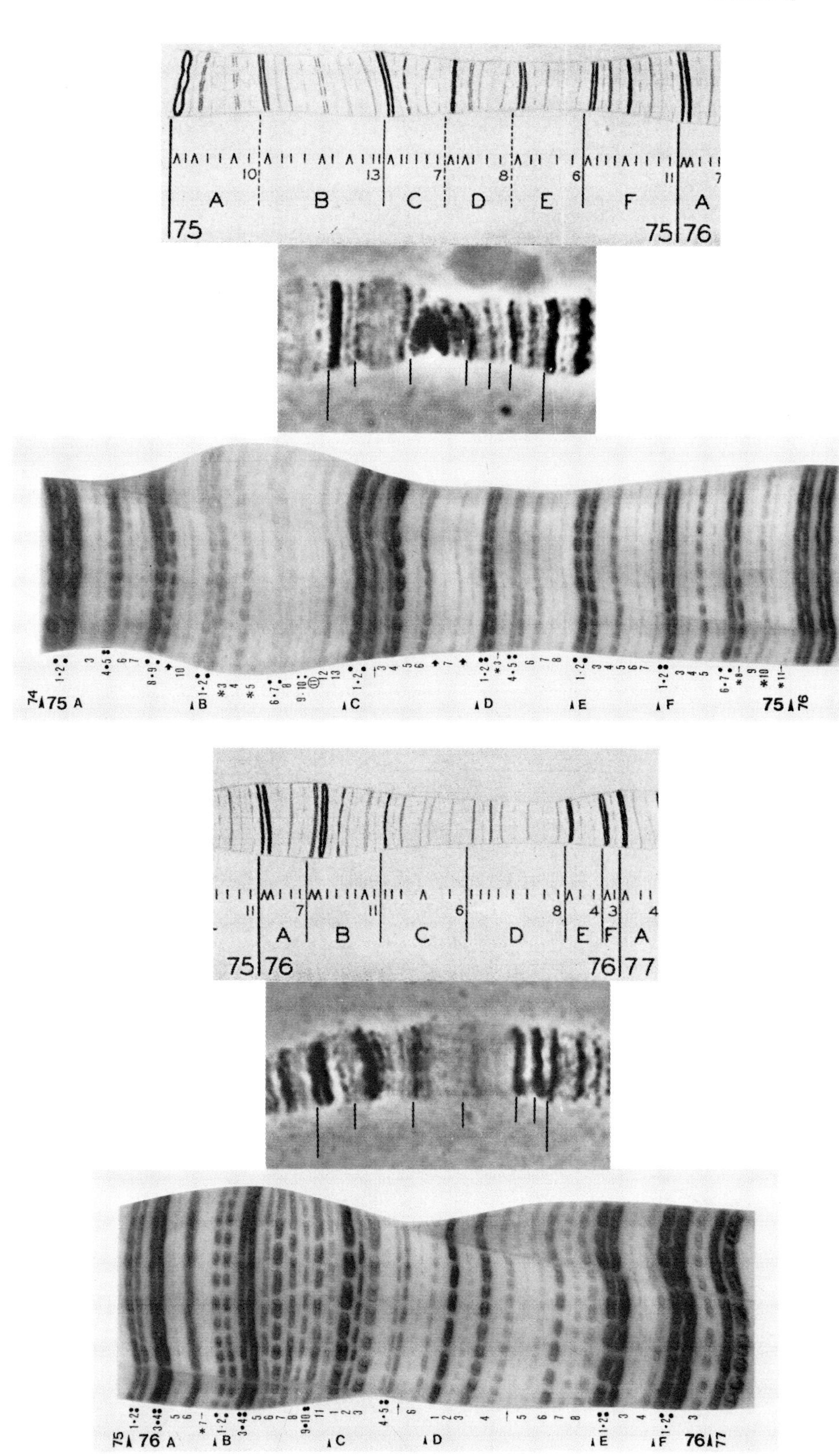

FIGURE 1 (continued)

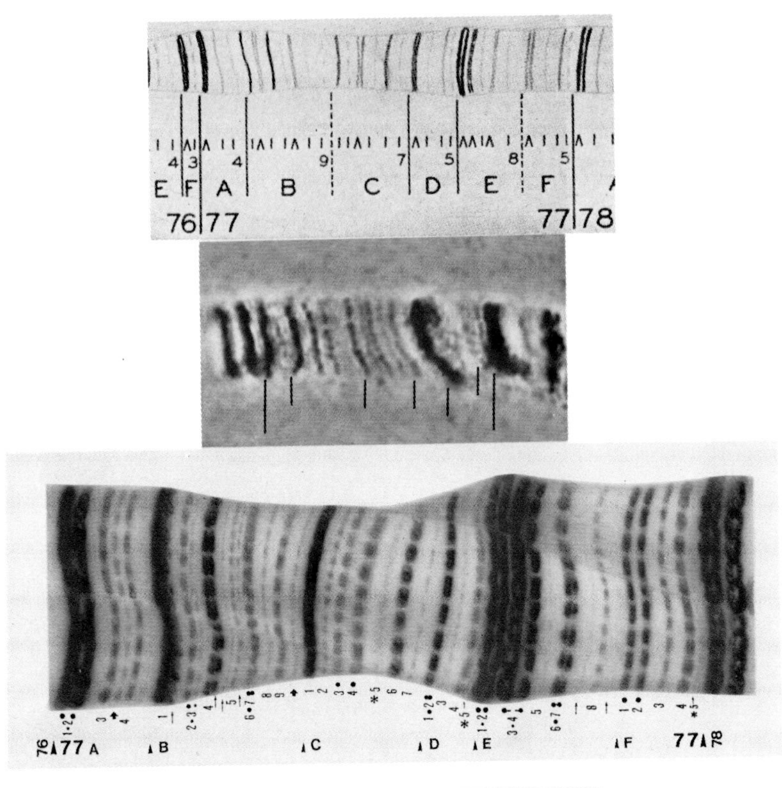

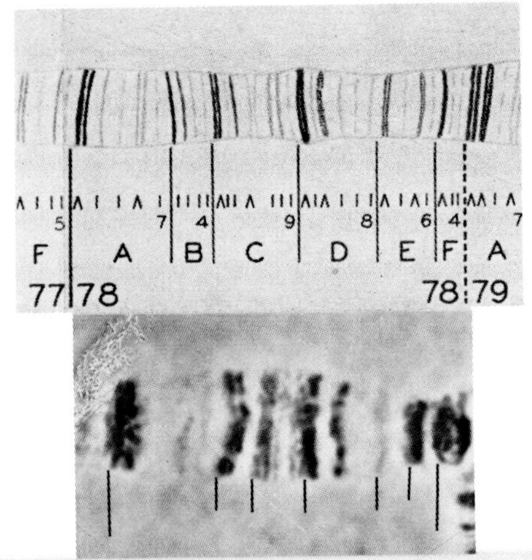

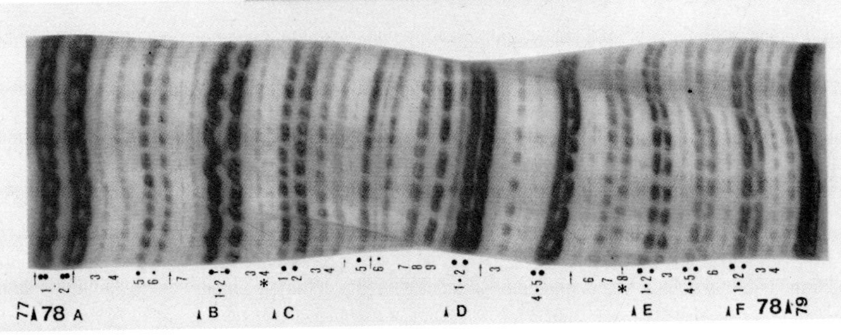

FIGURE 1 (continued)

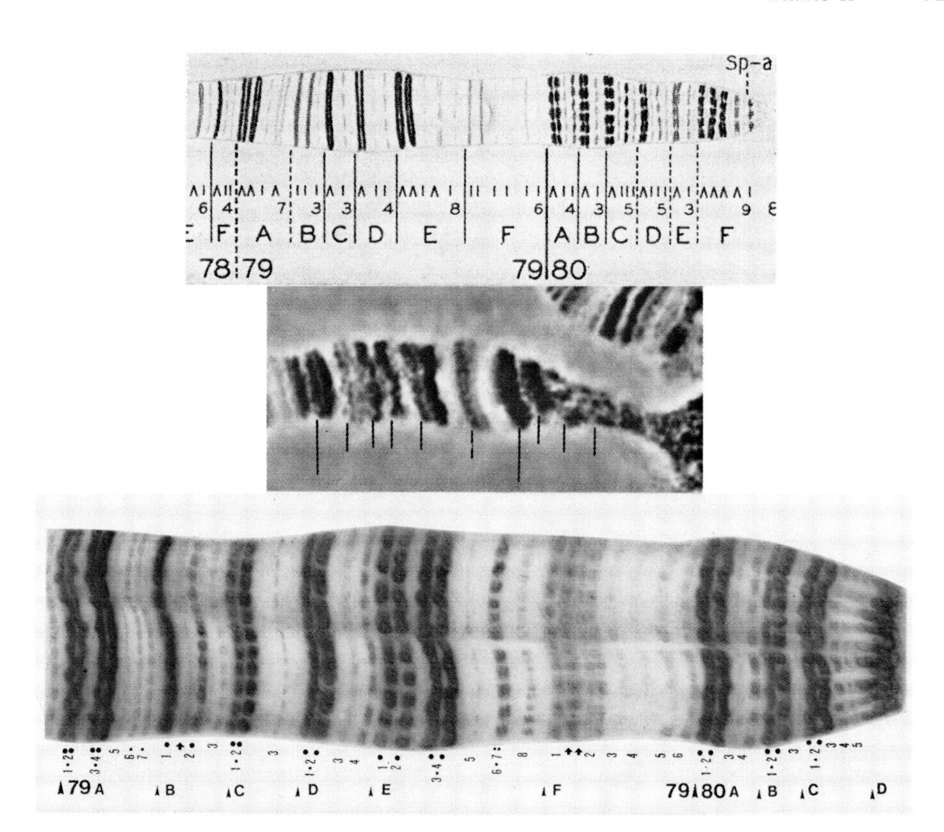

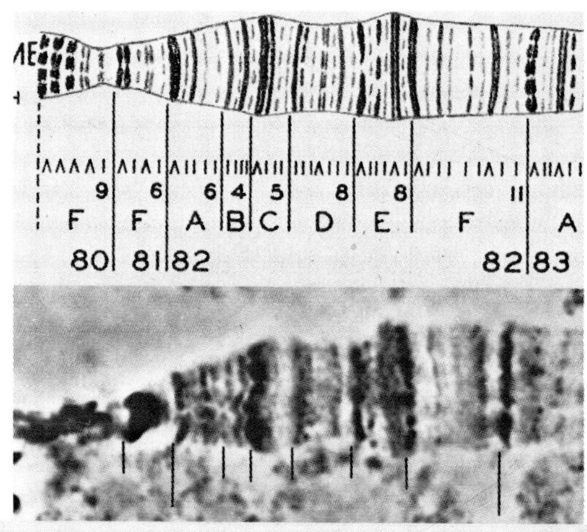

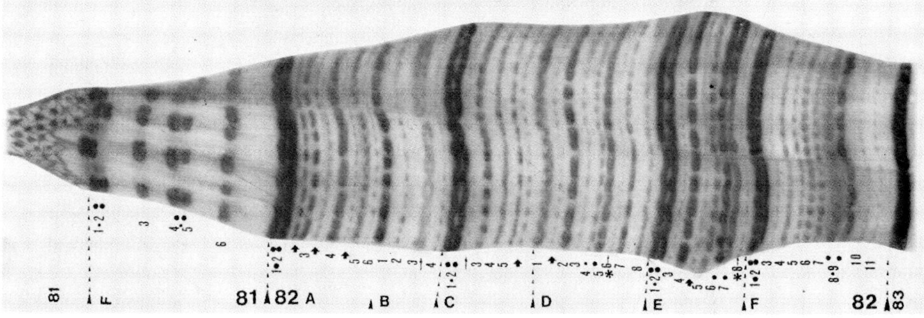

FIGURE 1 (continued)

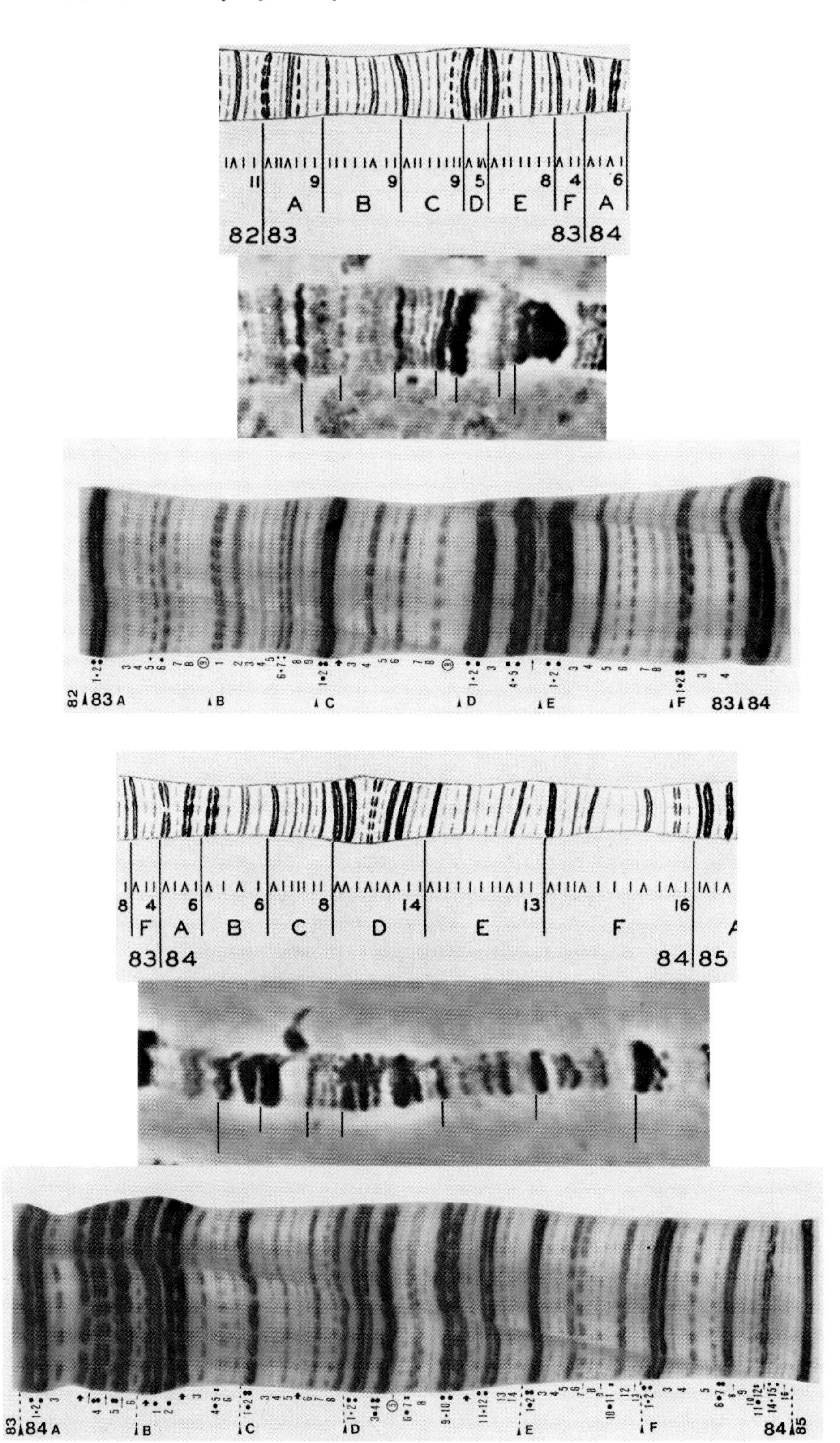

FIGURE 1 (continued)

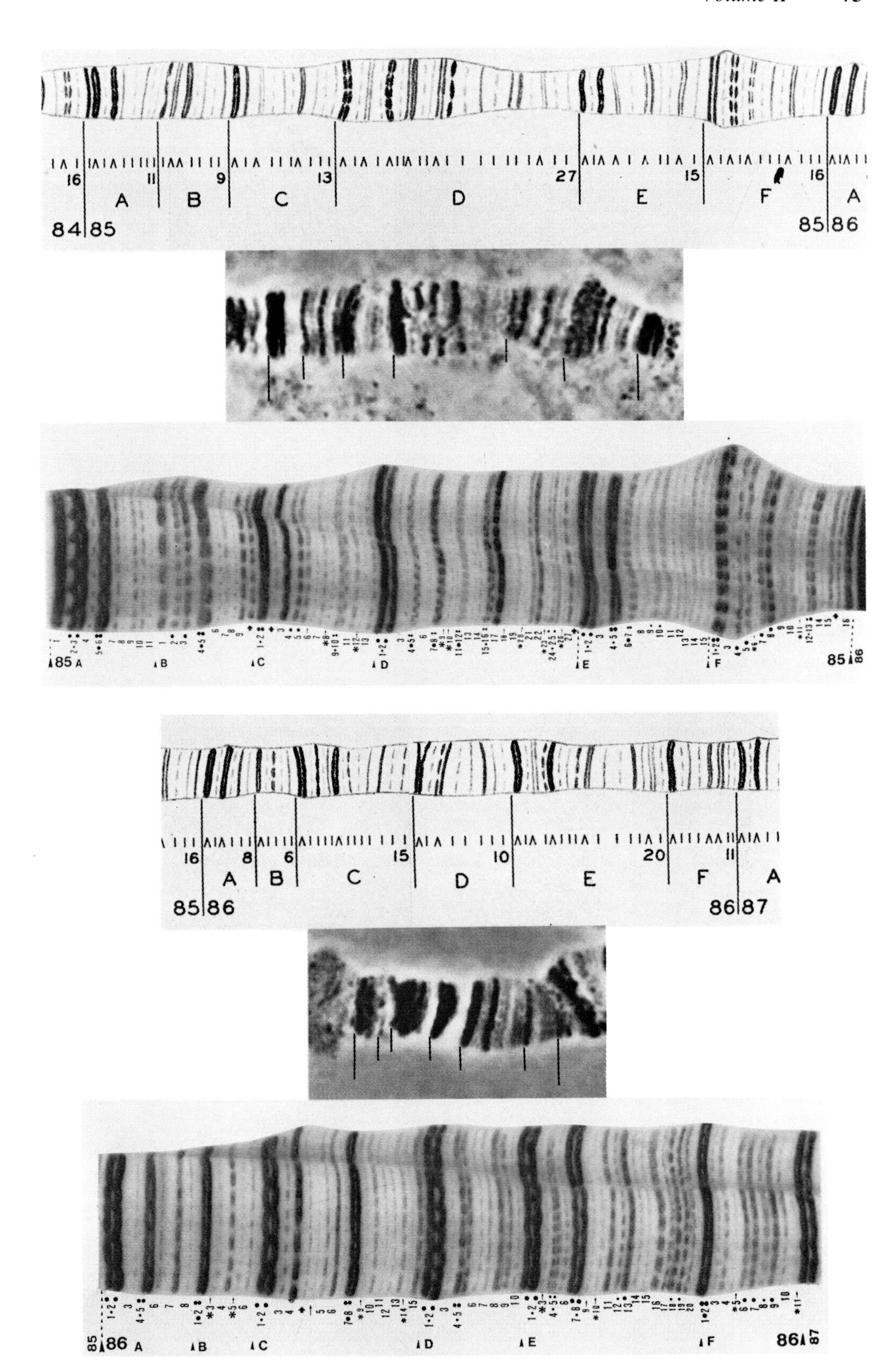

FIGURE 1 (continued)

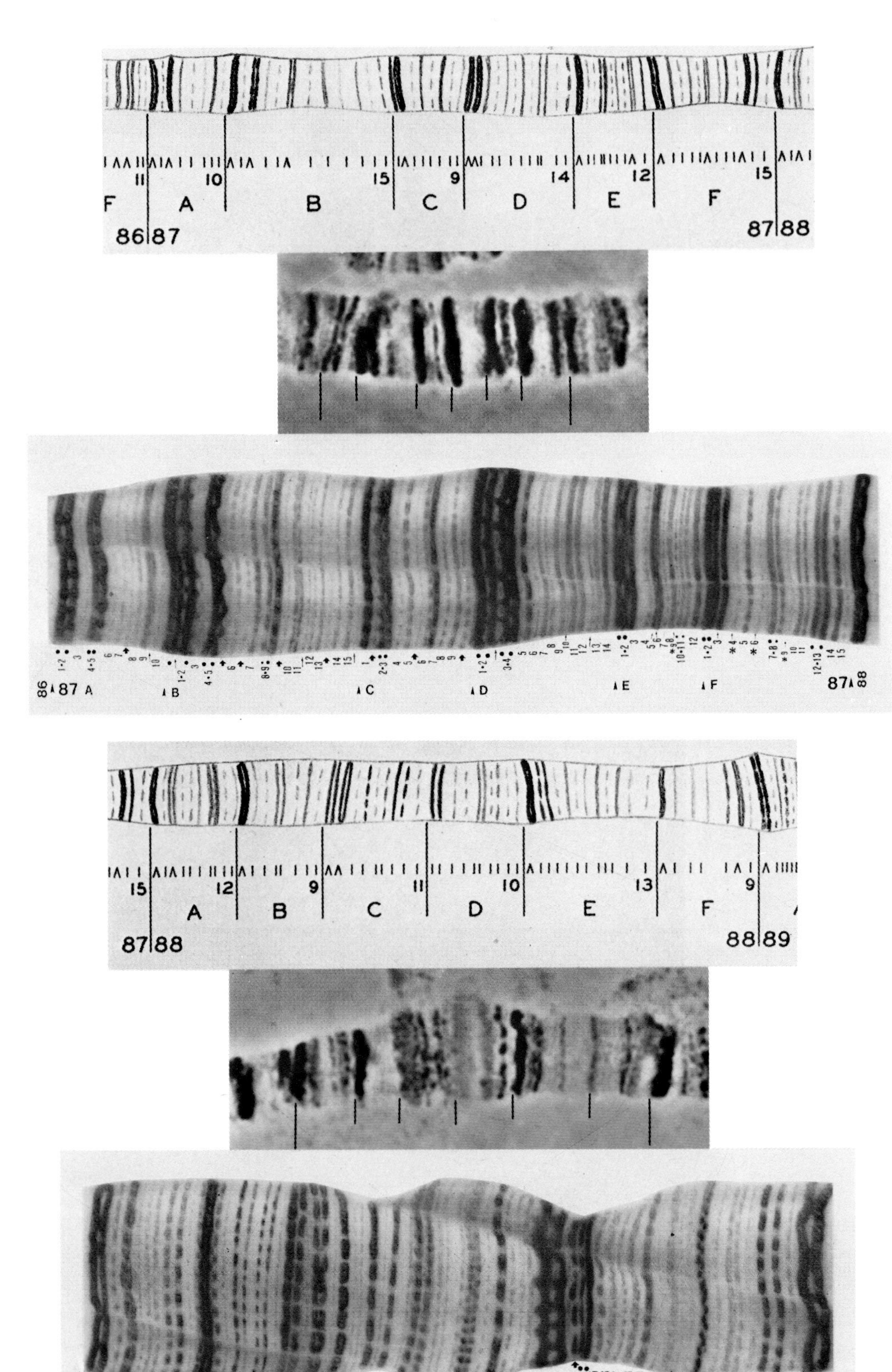

FIGURE 1 (continued)

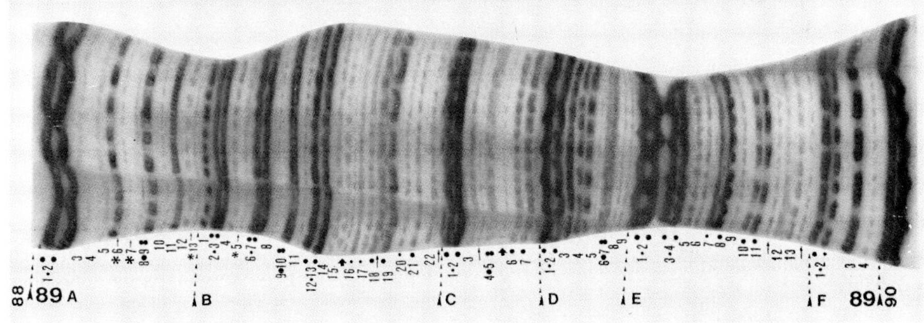

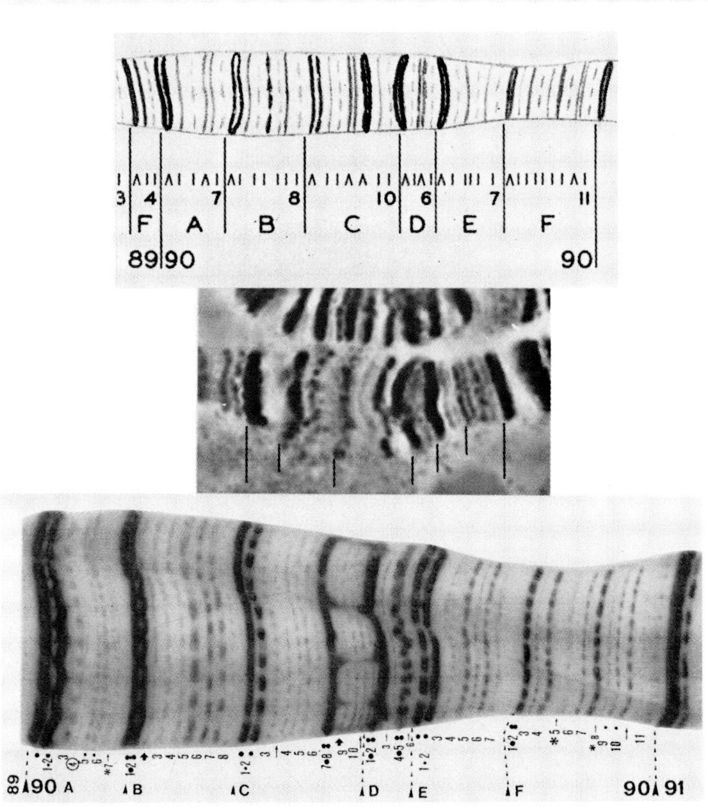

FIGURE 1 (continued)

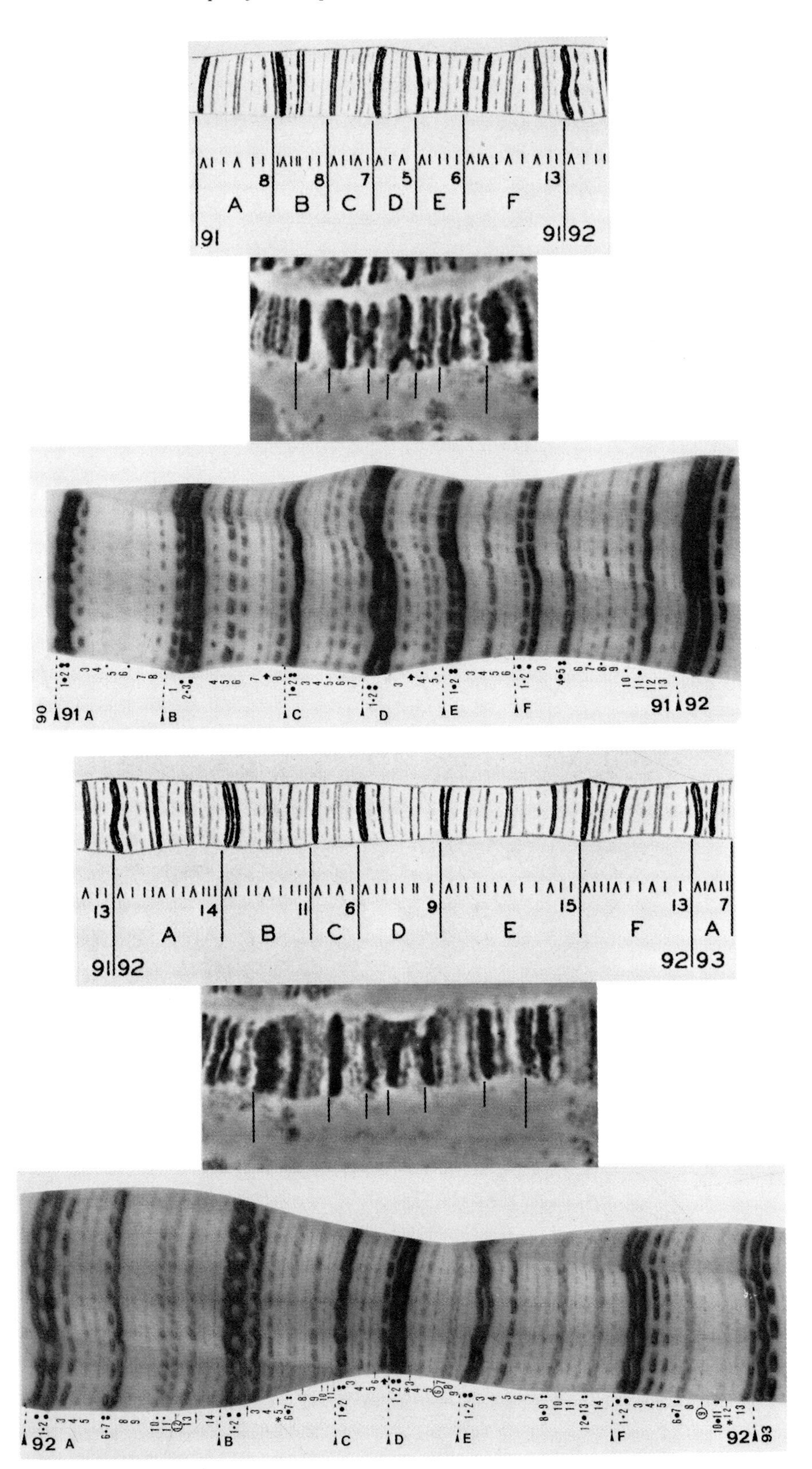

FIGURE 1 (continued)

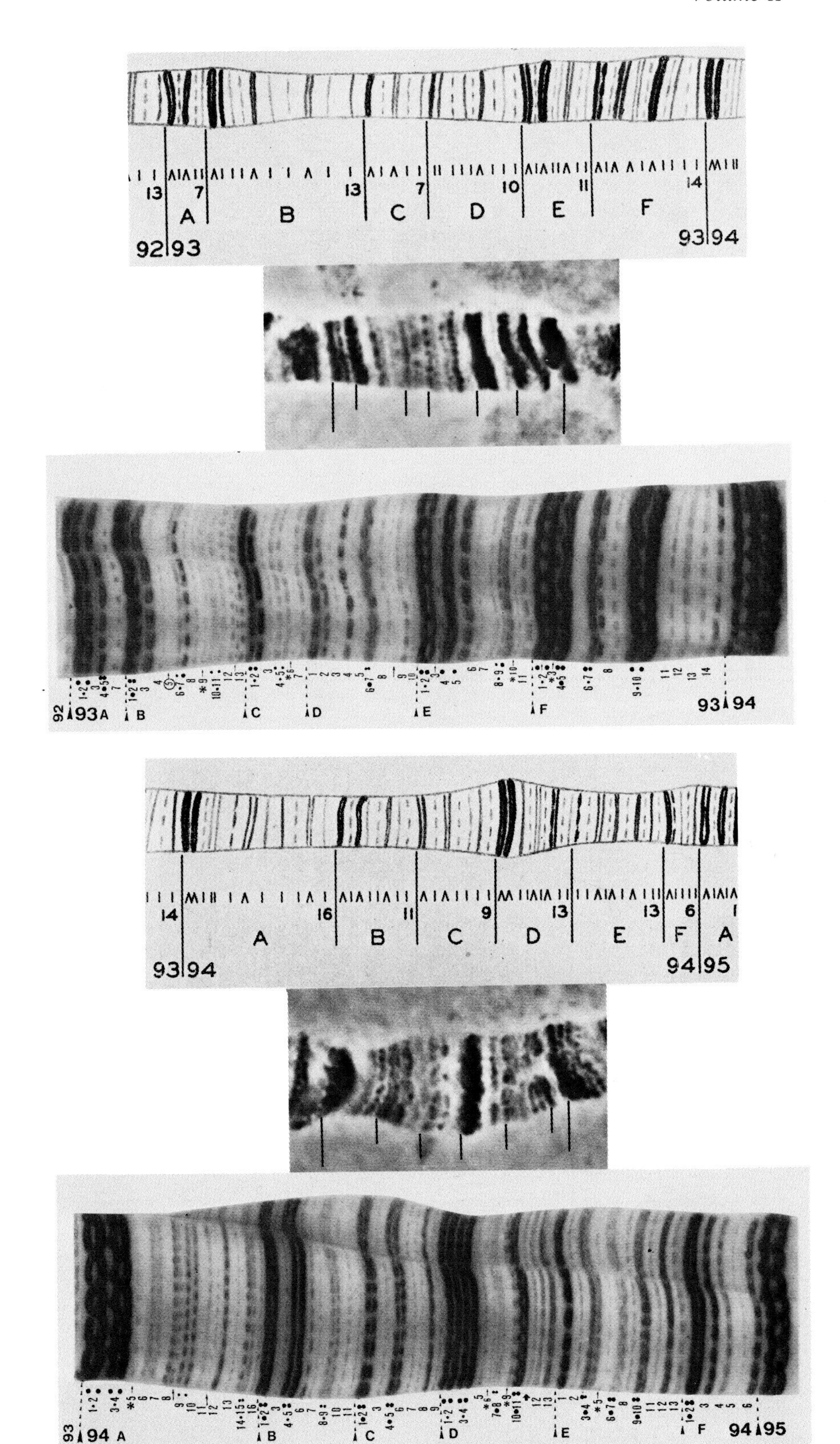

FIGURE 1 (continued)

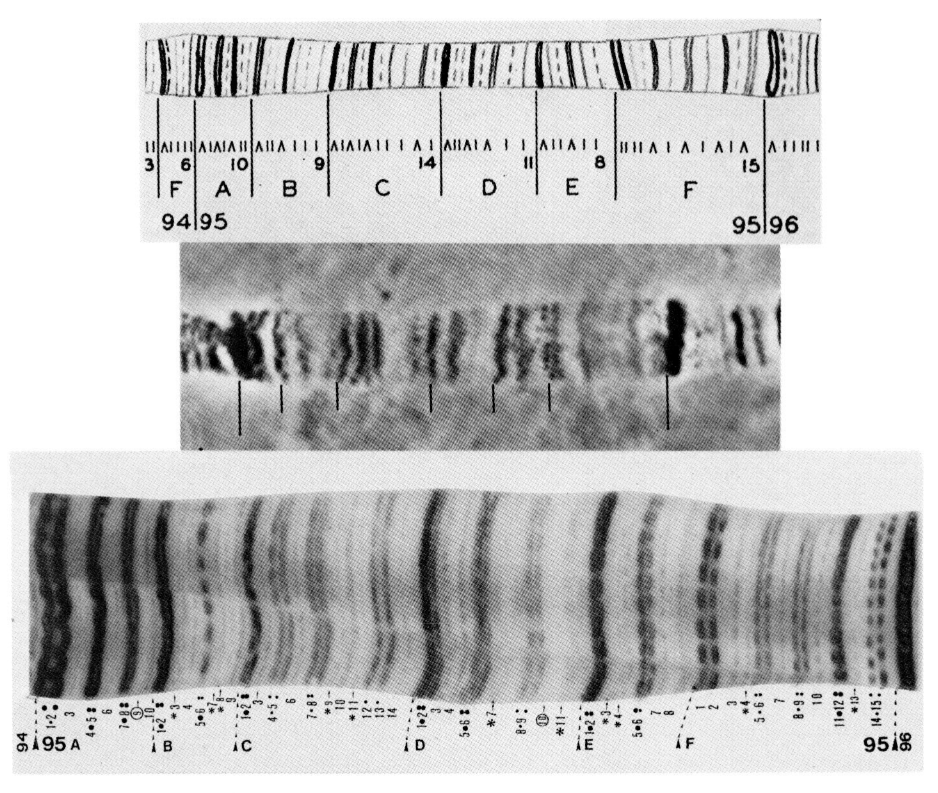

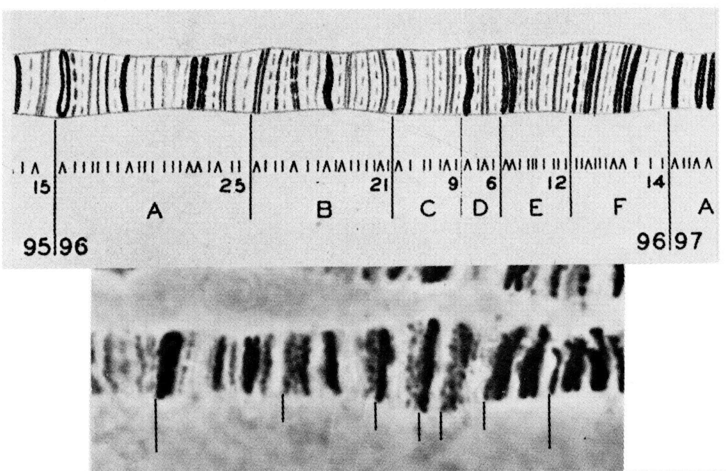

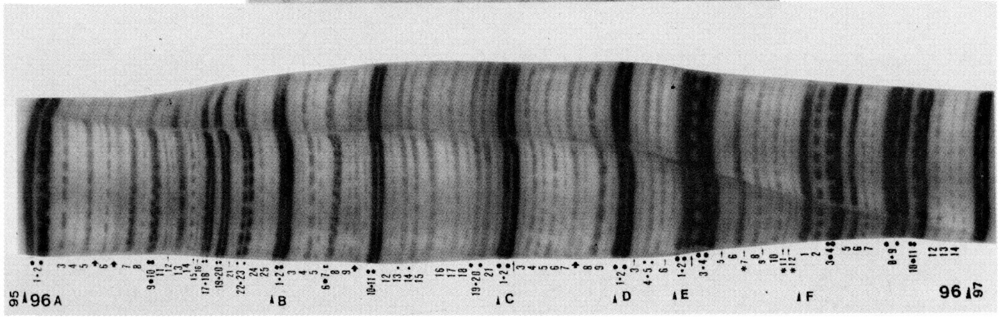

FIGURE 1 (continued)

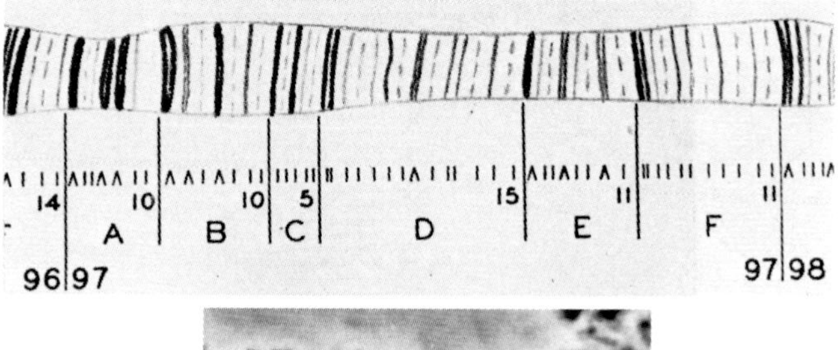

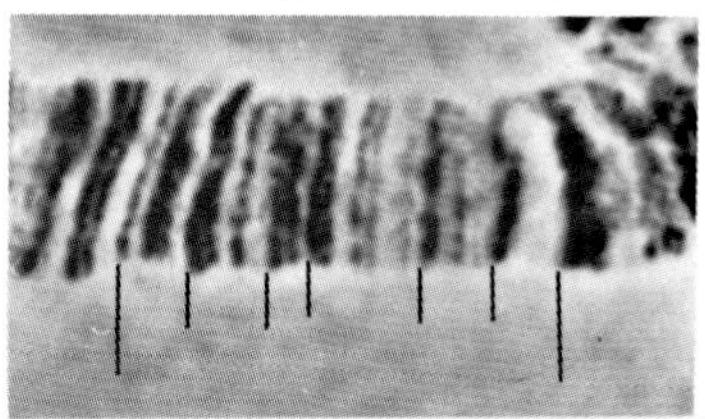

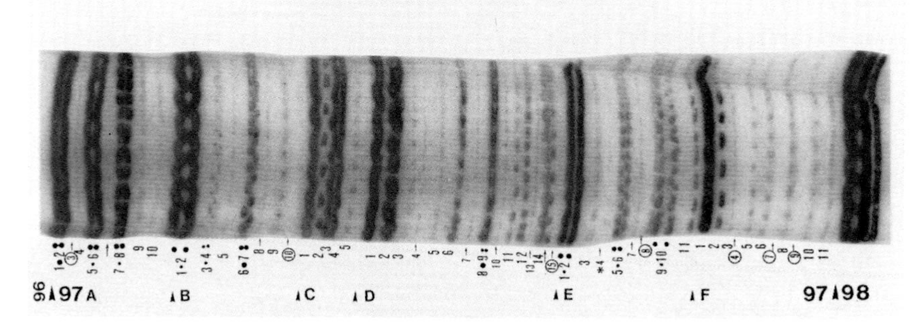

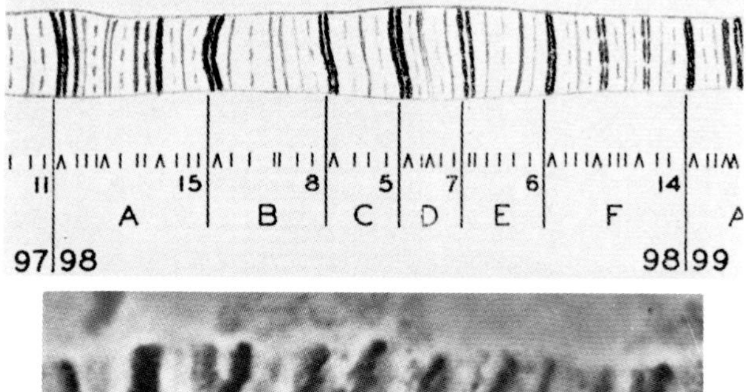

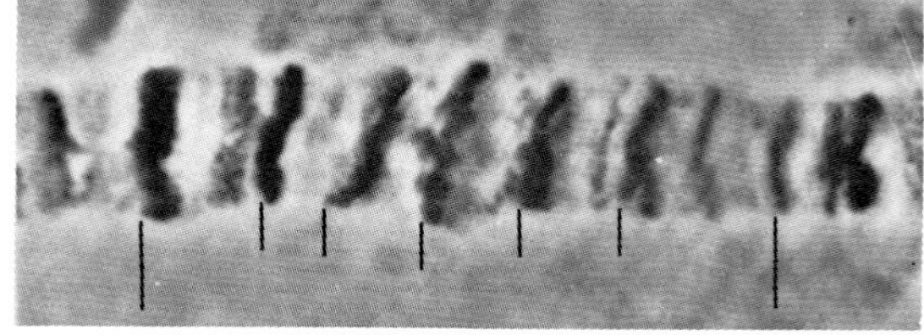

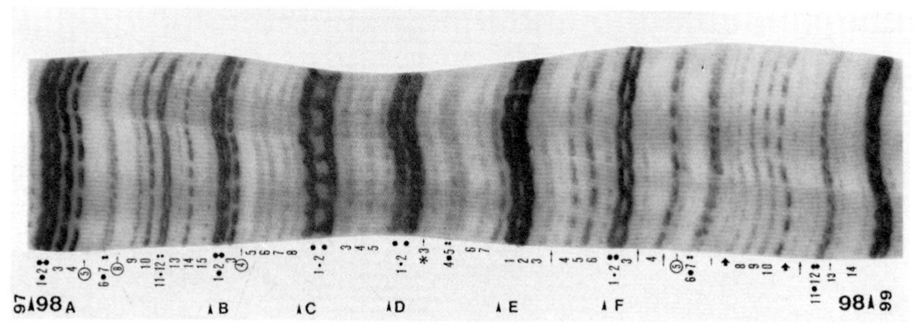

FIGURE 1 (continued)

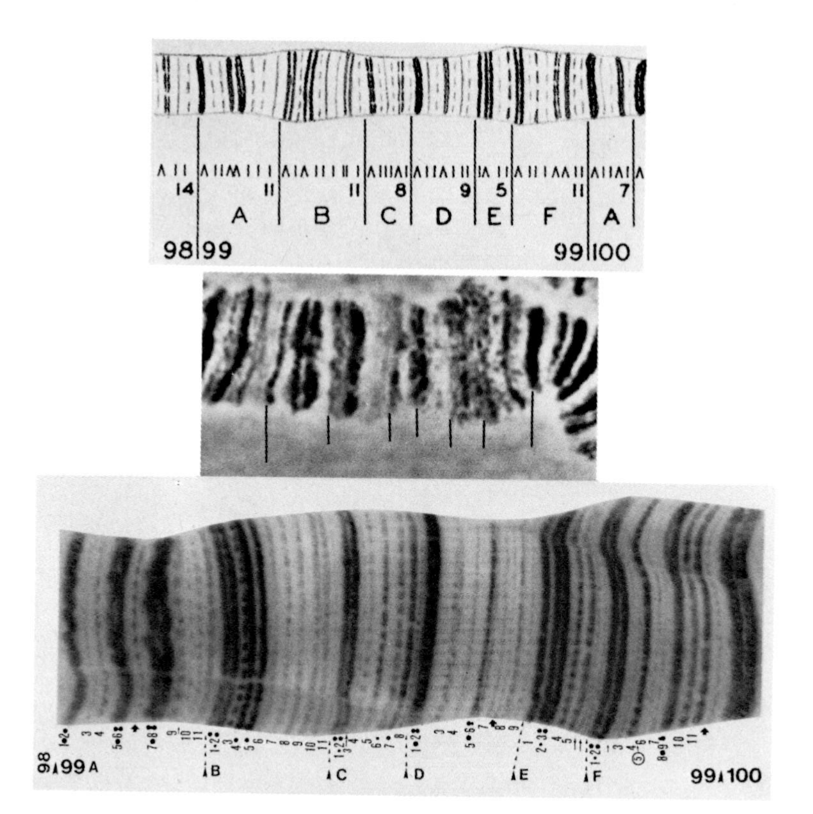

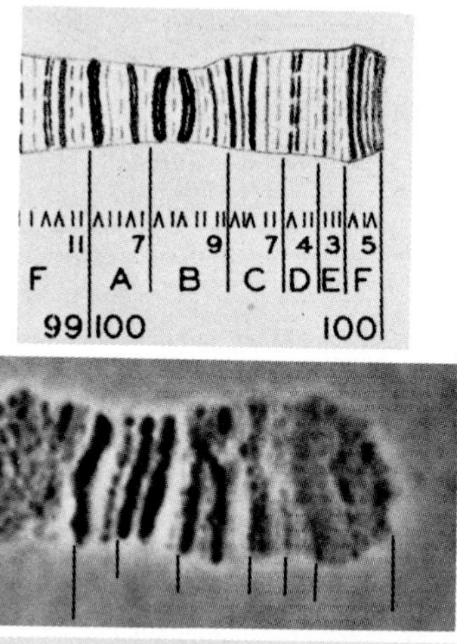

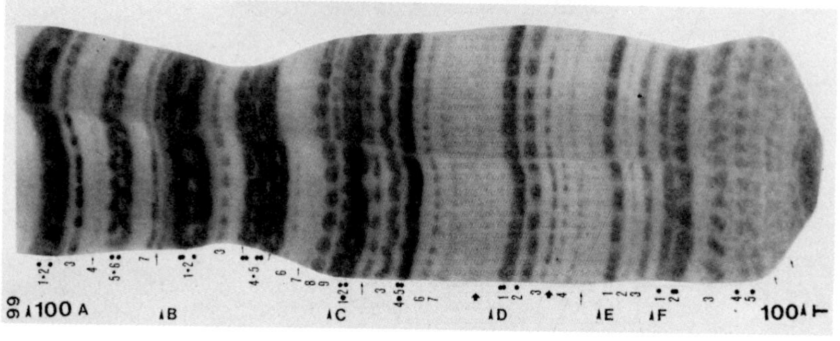

FIGURE 1 (continued)

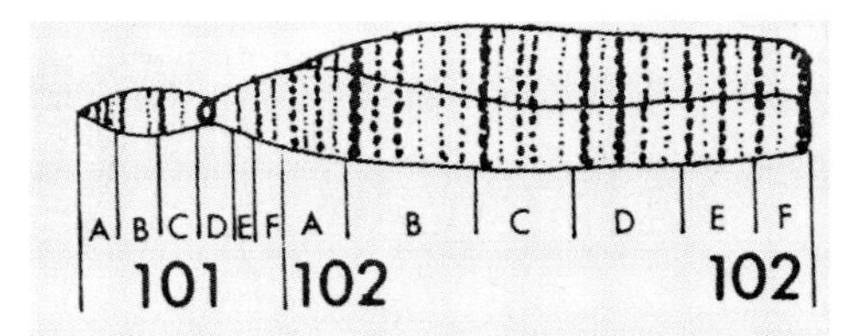

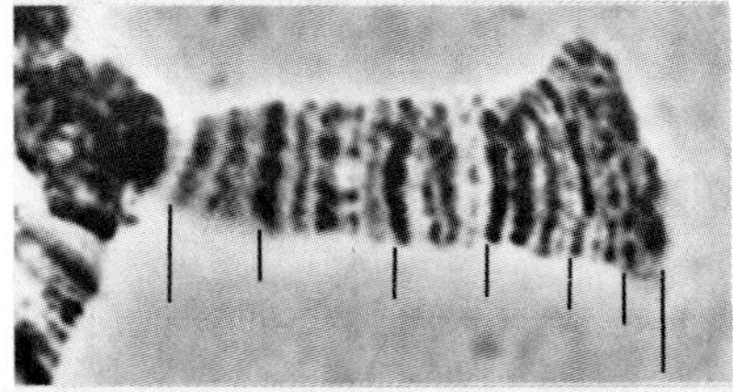

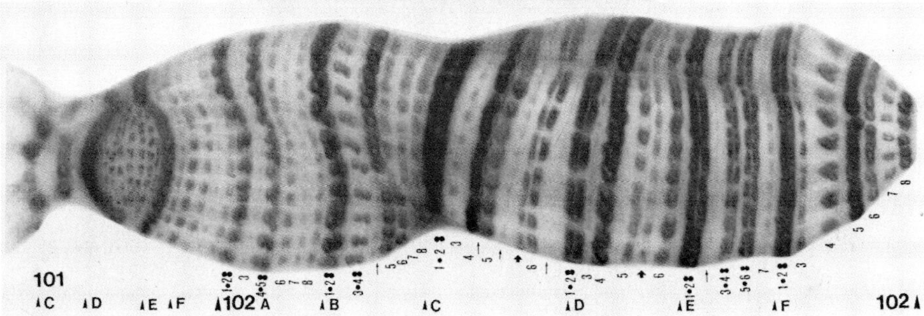

1	DIVISION BORDER
2 A	Subdivision border
1	A single band, identified and numbered according to the revised reference maps of BRIDGES
2 • 3 •	A pair of single bands in the EM, although depicted as a doublet in the reference maps of BRIDGES
4 · 5	A clear double band also in the EM
6 • 7	Occasionally a clear doublet also in the EM, and thus interpreted as double band in the present EM map
8 • 9	A BRIDGES´ doublet, but mostly like a single band in the EM, and thus interpreted as singlet in the EM map
10 →	A BRIDGES´ band sometimes difficult to detect in the EM of thin sections analyzed upto now
*11 →	A BRIDGES´ band only occasionally detected in thin section EM or otherwise unclear for exact characterization in EM
(12)	A BRIDGES´ band not found in the thin sections analyzed for the present EM map
! (13 14	Correspondence of bands to the reference maps not clear
➡	A new band detectable in the EM in all or most thin sections
→	A new band occasionally detectable in thin sections
15 →	A new band close to numbered BRIDGES´ band showing like doublet in the EM

FIGURE 1 (continued)

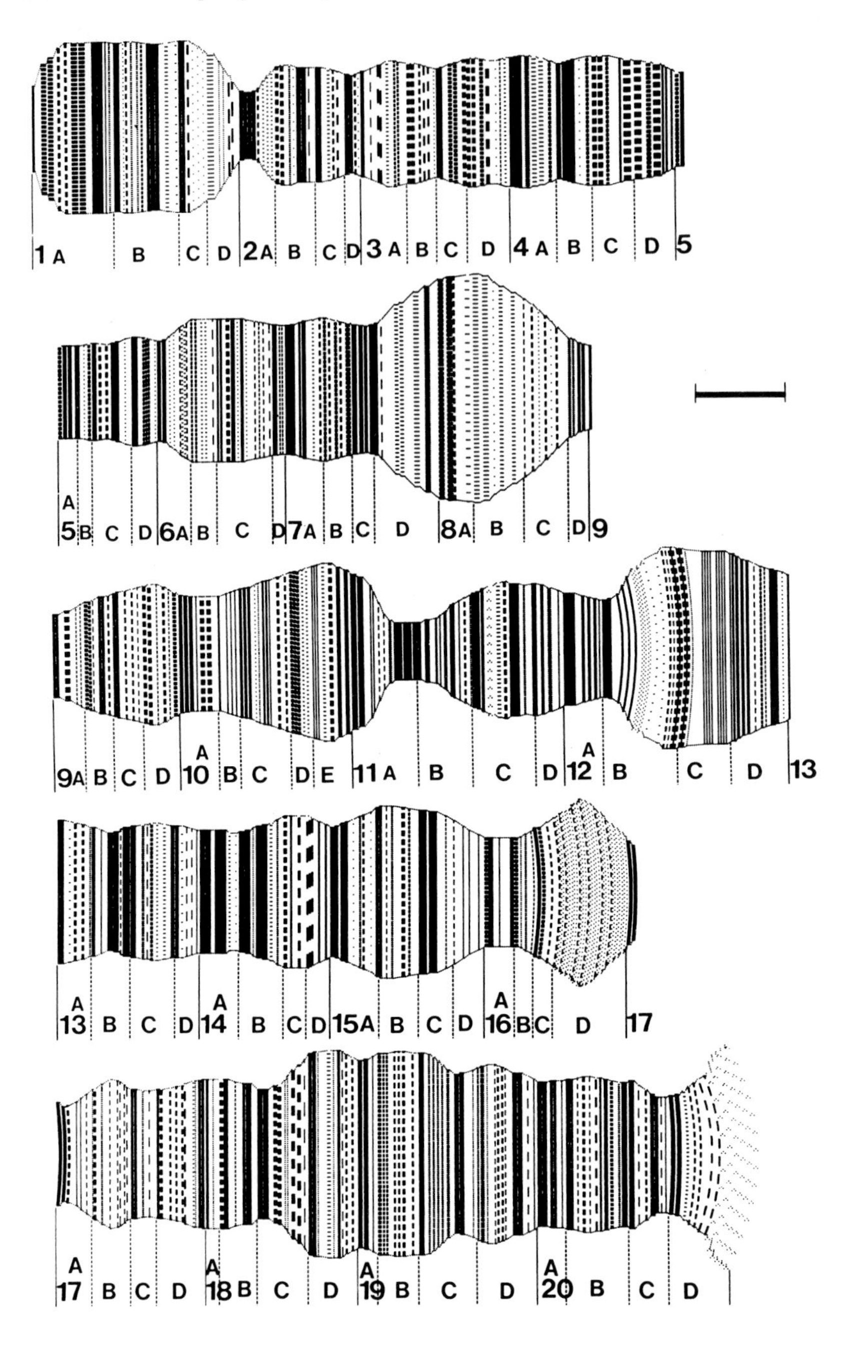

FIGURE 2. A computerized map of the X chromosome of *Drosophila hydei*. The map has been plotted by using the average dimensions of bands and interbands calculated on the basis of a number of measurements made from the electron micrographs of SSP preparations. Apparently, due to the whole mount spreading method used for the preparation of chromosomes for the electron microcopy the regions of heavy bands are more constricted, and contrariwise, the regions formed by tiny bands are more spread than in the normal squash preparations. The spreading increases the diameter of chromosomes remarkably as is shown by the scale bar representing 10 μm. (Courtesy of Dr. W.-E. Kalisch.)

Table 1
COMBINED CATALOGS OF BAND NUMBERS, DNA CONTENTS, AVAILABLE CLONES AND TRANSFORMED STRAINS AS WELL AS SOME EXAMPLES OF LOCALIZED GENES IN THE SALIVARY GLAND GENOME OF *DROSOPHILA MELANOGASTER*

X chromosome

a (D)	b (SD)	c (BN)	d (DNA-C)	e (TDF)	I (CL)	II (TR)	III (Localized genes:) (with cross over/chromosomal site)
1	A	10	225		##		ac, arth, cin, ewg, dmd l(1) J1, y (0.0/1A5-8)
	B	16	367		##	**	Hw (0.0/1B1-2); cp 70 (0.0/1B1-2B); sc (0.0/1B3); svr (0.0/1B5-6) su(s) (0.1/1B11-13);
	C	7	176		#	*	su(b) (0.1/1B4-C4); arm (0.2/1B4-E4); M(1) Bld (0.1/1B11-13); om (0.1/1C1)
	D	4	78			*	su (w^{a})) (0.1/1D1-E); sta (0.3/1D3-2B)
	E	5	137				fs(1) N (?/1E1-2A)
	F	7	142			**	dor (0.3/1F1-2A2); tw (0.4/1C5-2C10)
1	A-F	49	1205	1.14			
2	A	7	127			**	
	B	21	450		##	**	l(1)npr1 (0.5?/2B5); fmf (0.5/2B7-12); cripA (?/2B11);
	C	11	235		#	*	
	D	8	176		#		Pdg (0.63/2D3-4); pn (0.8/2D5-6)
	E	3	88		##		kz (0.9/2E1-F6); crn (0.9/ 2E2-F1); fs(1)K10 (0.5?/2E2-F1)
	F	6	176		##		
2	A-F	56	1252	1.03			
3	A	14	313		#	*	gt (?/3A1); z (1.0/3A3); l(1) zw1,l(1) zw4, l(1) zw8 (1.1/3A5-8)
	B	7	171		##	*	l(1) zw2 (1.2/3A7-B1); l(1) zw6 (1.3/3A7-B1); l(1) zw5, l(1) zw7 (1.4/3B3-C2); fs(1)Ya, Yb (?/3B4-6);
	C	14	464		##		crm (1.48/3C1); w (1.5/3C2); rst (1.7/3C4); Co (ca.3/3C5-6); Ax, fa, N, nd, spl (3.0/3C7); Sgs4 (3.6/3C11-12)

Table 1 (continued)
COMBINED CATALOGS OF BAND NUMBERS, DNA CONTENTS, AVAILABLE CLONES AND TRANSFORMED STRAINS AS WELL AS SOME EXAMPLES OF LOCALIZED GENES IN THE SALIVARY GLAND GENOME OF *DROSOPHILA MELANOGASTER*

X chromosome

a (D)	b (SD)	c (BN)	d (DNA-C)	e (TDF)	I (CL)	II (TR)	III (Localized genes:) (with cross over/chromosomal site)
	D	8	152		#		A(4.5/3C-D) dm (4.6/3C9-D2); dnc (4.6/3D4)
	E	9	132				M(1)3E (5.0/3E3-4)
	F	9	171			*	ec(5.5/3F1-2); cho(5.5/3F)
3	A-F	61	1403	1.06			
4	A	6	152				
	B	8	215		##	*	M(1)4BC (6.8/4B5-C6)
	C	20	391		##	*	bi (6.9/4C7-D2); hnt (7.0/4B1-C15); peb (7.3/4C7-D2)
	D	10	200			*	rb (7.5/4C8-D1)
	E	3	98				rg (11.0/4E1-3)
	F	12	284		##		cx (13.6/4F1-5D2); cv (13.7/4F1-5D2)
4	A-F	59	1340	1.05			
5	A	15	323		##		
	B	7	176		#		
	C	11	210		##		Act 5C (14.0/5C3-4); l(1)5CD (15.1/5C6-D6)
	D	8	274		#		M (1)30(14?/5D3-6A2)
	E	8	132		#		vs (16.3/503?-6A2)
	F	9	122		##		Fum (18.0/5F1-?)
5	A-F	58	1237	0.99			
6	A	4	137			*	dx (17.0), shf (17.9), cm (18.9/6A3-F9)
	B	4	78			*	scp (19.3/6A3-6F9)
	C	19	386				
	D	8	181				
	E	6	171				
	F	9	186		##	*	
6	A-F	50	1139	1.05			
7	A	9	196		#		
	B	12	328		#		ct (20.0/7B3-4); bis (20.1/7B6-7); ag (20.7/7B4-C1)
	C	10	249				dec1 (20.8/7C)
	D	21	435		##	*	fs(1)1163, l(1)93p, fs(1)h, clw, vtw (ca.21/7D1-6)
	E	9	259		##		s36, s38, oc (23.1/7E11-F2); dec2 (23.1/7D10-8A5);

Table 1 (continued)
COMBINED CATALOGS OF BAND NUMBERS, DNA CONTENTS, AVAILABLE CLONES AND TRANSFORMED STRAINS AS WELL AS SOME EXAMPLES OF LOCALIZED GENES IN THE SALIVARY GLAND GENOME OF *DROSOPHILA MELANOGASTER*

X chromosome

a (D)	b (SD)	c (BN)	d (DNA-C)	e (TDF)	I (CL)	II (TR)	III (Localized genes:) (with cross over/chromosomal site)
	F	10	205		##		gg (23.1/7F10); ptg (23.2), dd (24.3/7C4-8C2); tbd (25.0/7C5?-8C1); con (27.1/7C4-8C2)
7	A-F	71	1672	1.09			
8	A	6	142		#		
	B	8	161		#	*	
	C	15	269		#	*	t (27.5/8C3-17);
	D	14	269		#		amx, lz, su(r) (27.7/8D4-E2); Hex-A (29.2?/8D4-E2)
	E	11	230		#	*	gmp (?/8E-9D); dvr(28.1/8D8-9)
	F	10	156		##		Yp1, Yp2(ca.29/8F-9A)
8	A-F	64	1227	0.89			
9	A	7	318		##	*	btd (31.0/8A5-9A1)
	B	14	342			**	flp (31/9B1-10A1)
	C	6	147			*	
	D	4	117		#	*	
	E	9	230			**	gua1(31?/), ras (32.8/9E1-3); pur1(?/9E1-3); fliK (?/9E4-F5)
	F	12	308				fs(1)BP (32.67/9F12)
9	A-F	52	1462	1.30			
10	A	9	284		##		v (33.0/10A1); fs(1)10A (33.52/10A8)
	B	16	504		##	*	dsh (34.05/10B4-C2)
	C	9	254		##	*	RpII215 (35.7/10C1-2)
	D	8	137		#	*	
	E	7	127		#		m(36.1/10E1-2); dy(36.2/10E2); And (36.2/10E2-4)
	F	12	196		##		Hsp10F (?/10F1); ftd (36/10F1-10); Flu (37.0/10F7-11D1)
10	A-F	61	1502	1.14			
11	A	12	298		##		gd (36.78/11A1-7); fw (36.85/11A4-); mfd (38?/11A6-7); Lsp 1a (39.5/11A7-B9); clv2 (42?/11A7-8)
	B	17	308				

Table 1 (continued)
COMBINED CATALOGS OF BAND NUMBERS, DNA CONTENTS, AVAILABLE CLONES AND TRANSFORMED STRAINS AS WELL AS SOME EXAMPLES OF LOCALIZED GENES IN THE SALIVARY GLAND GENOME OF *DROSOPHILA MELANOGASTER*

X chromosome

a (D)	b (SD)	c (BN)	d (DNA-C)	e (TDF)	I (CL)	II (TR)	III (Localized genes:) (with cross over/chromosomal site)
	C	3	88				
	D	10	230		#		wy (41.9/11D-E)
	E	9	196				l(1)ts403 (42.0/11E-F)
	F	7	127				Gpt (42.6/11F1-12A2)
11	A-F	58	1247	0.99			
12	A	9	259			*	int,up (43.5/12A1-7); ben (?/12A6-D3);
	B	11	235		#	**	Yp3, Yp3R (ca.44/12B-C)
	C	6	122		#	**	
	D	6	117		##	*	g (44.4/12B9-C7)
	E	12	274		##	**	fs(1)29 (44.5/12E1-F1)
	F	5	147		##		eag (50?/12F6-13A4)
12	A-F	49	1154	1.09	#		
13	A	9	240			*	bas (49.5?/13A1)
	B	9	210			*	pl (47.9/13B2-F17)
	C	6	147			*	sd (51.5/13B2-F17)
	D	5	93			*	
	E	14	303		#	*	C-myb (50?/13EF)
	F	13	264		##	*	exd (54?/13F-14B1)
13	A-F	56	1257	1.04			
14	A	8	205			*	
	B	13	293		#		eas (53.5/14B1-13); bss (54.0/14B1-13); U5 (? /14B)
	C	5	108		#		
	D	3	88		#		l(1)15A (53?/14D1-15A1)
	E	4	78				
	F	5	127			*	Had (54.4/14D-F)
14	A-F	38	899	1.10			
15	A	9	235		#		r (55.3/15A1); baz (56.7/15A5-16A2); Su(b) (55.5/15A1-)
	B	4	83		##		M(1)o (56.6/15B1-E7)
	C	4	78				
	D	4	108			*	
	E	4	98			**	
	F	6	137		##		f (56.7/15F1-5); gnd (58/15-16)
15	A-F	31	739	1.10			

Table 1 (continued)
COMBINED CATALOGS OF BAND NUMBERS, DNA CONTENTS, AVAILABLE CLONES AND TRANSFORMED STRAINS AS WELL AS SOME EXAMPLES OF LOCALIZED GENES IN THE SALIVARY GLAND GENOME OF *DROSOPHILA MELANOGASTER*

X chromosome

a (D)	b (SD)	c (BN)	d (DNA-C)	e (TDF)	I (CL)	II (TR)	III (Localized genes:) (with cross over/chromosomal site)
16	A	3	181				B (57.0/16A1,2); E(B) (57.3/16A5-6)
	B	10	205		#	*	
	C	9	176			**	
	D	8	142			*	
	E	4	83			*	
	F	6	210		##	*	
16	A-F	40	997	1.15	#		
17	A	10	362		#		Bx (59.4/17A-C)
	B	5	152		#		hdp-a,hdp-b(59.4/17C2-3)
	C	5	161			*	
	D	5	132			*	fu (59.5?/17D-E)
	E	5	117			*	
	F	2	49				
17	A-F	32	973	1.4			
18	A	8	181			**	sby, pph, srb (60.8/18A4-B8); smd (61.1), coc (61.5)/(18A4-B8); gfA (61.62/18A5-D12)
	B	10	142				
	C	7	210		#		
	D	15	225		##	**	car (62.5/18D1-2); Zw (62.9/18D)
	E	4	78		##		
	F	5	132				
18	A-F	49	968	0.91			
19	A	6	156			*	amn (60?/19A1-2)
	B	3	68				
	C	5	156				
	D	3	98				ma-l (64.8/19D1-3)
	E	14	254		##	**	Cg19-20 (65?/19E-20B); sdby (ca.65/19E-20B)
	F	6	122		##	*	
19	A-F	37	854	1.07			
20	A	4	127		##	**	fog (65/20AB)
	B	2	98		#		
	C	2	98			*	bb (66.0/20C1-2);
	D	?	?			*	su(f) (65.9?/20D-F)
	E	?	?				
	F	?	?				

Table 1 (continued)
COMBINED CATALOGS OF BAND NUMBERS, DNA CONTENTS, AVAILABLE CLONES AND TRANSFORMED STRAINS AS WELL AS SOME EXAMPLES OF LOCALIZED GENES IN THE SALIVARY GLAND GENOME OF *DROSOPHILA MELANOGASTER*

X chromosome

a (D)	b (SD)	c (BN)	d (DNA-C)	e (TDF)	I (CL)	II (TR)	III (Localized genes:) (with cross over/chromosomal site)
20	A-F	8?	323?				

2L chromosome

a (D)	b (SD)	c (BN)	d (DNA-C)	e (TDF)	I (CL)	II (TR)	III (Localized genes:) (with cross over/chromosomal site)
21	A	5	127			##	l(2) gl (0.0/21A-C); net (0.0/21A1-C1)
	B	15	215		#	*	
	C	13	308		#		M(2)21C1-2(0.0/21C1-2); al 0.01/21C1-2);
	D	7	205		##	**	ds (0.3?/21D1-2);
	E	7	191		#	*	ast (1.3/21E1-2); S (1.5/21E2); Lsp1B(1.9/21D-22A)
	F	4	98		#		
21	A-F	51	1144	1.04			
22	A	14	284		#	*	shr (2.3/22A3-B1)
	B	11	249		##	*	Got-2 (3.0/22B1-4)
	C	3	88		#		
	D	9	127		#		Su(S) (3.0/22D1-E1)
	E	4	88				ho (4.0/22E)
	F	6	152		##	*	fs(2)B (5.6/22-23); Pgk (5.9/22D-23E3); tRNA Tyr (?/22F1-6)
22	A-F	47	988	0.97			
23	A	7	152		#	**	tRNA Thr-lg (?/23A)
	B	10	147		#	**	
	C	8	112		#	*	
	D	8	132				U5 (?/23D)
	E	7	108		#		msl2 (9.0/23E-F6)
	F	4	108				
23	A-F	44	759	0.80			
24	A	5	108		#	*	
	B	2	54		#	*	
	C	8	152		#	*	
	D	8	176			*	ed(11.0)ft(12/24D2-F1); G (12.0/24D2-F1)
	E	7	166				M(2)2 (12.9/24E2-25A2)
	F	8	142				dp,dw-24F (13.0/24E3-25A)
24	A-F	38	798	0.97			

Table 1 (continued)
COMBINED CATALOGS OF BAND NUMBERS, DNA CONTENTS, AVAILABLE CLONES AND TRANSFORMED STRAINS AS WELL AS SOME EXAMPLES OF LOCALIZED GENES IN THE SALIVARY GLAND GENOME OF *DROSOPHILA MELANOGASTER*

2L chromosome

a (D)	b (SD)	c (BN)	d (DNA-C)	e (TDF)	I (CL)	II (TR)	III (Localized genes:) (with cross over/chromosomal site)
25	A	10	357			*	dp (13.0/25A1-2);
						**	E(var)7 (?/25A)
	B	11	259		#		sgs1 (13.9/25A3-D2)
	C	12	274		##	**	Cg 25C (ca.15/25C)
	D	7	127		##		inaB (?/25CD-28B)
	E	6	112				cl (18.83/25E1-2)
	F	7	152		#	*	Gpdh,Gdt3 (20.5/25F3-)
25	A-F	53	1281	1.12			
26	A	10	186		##		Gal (ca.20?/26A4-B3)
	B	13	210		#	**	Kf2 (?/25A-27E)
	C	4	103				
	D	11	161				ee (18?/26D1-2)
	E	6	98				
	F	8	132				
26	A-F	52	890	0.79			
27	A	2	68			*	
	B	4	122			*	
	C	8	176		##	**	Gart (?/27C); spd (21.9/27C1-28C1); Sp (22.0/27C-28C); ade3 (?/27C)
	D	6	117		#		
	E	9	249				
	F	7	161		#		
27	A-F	36	893	1.15			
28	A	5	127		#	*	
	B	3	68				
	C	8	127		##		C-src2 (?/28C)
	D	11	269		#	*	tRNA Gly-3 (?/28D)
	E	8	176				
	F	5	103				
28	A-F	40	870	1.01			
29	A	6	98		#		C-src2 (?/29A)
	B	6	83		#	*	
	C	6	103		#		
	D	6	93				
	E	6	117				
	F	11	196				
29	A-F	41	690	0.78			

Table 1 (continued)
COMBINED CATALOGS OF BAND NUMBERS, DNA CONTENTS, AVAILABLE CLONES AND TRANSFORMED STRAINS AS WELL AS SOME EXAMPLES OF LOCALIZED GENES IN THE SALIVARY GLAND GENOME OF *DROSOPHILA MELANOGASTER*

2L chromosome

a (D)	b (SD)	c (BN)	d (DNA-C)	e (TDF)	I (CL)	II (TR)	III (Localized genes:) (with cross over/chromosomal site)
30	A	10	196		#	**	
	B	14	259		##		
	C	9	171			**	
	D	3	117		#		
	E	5	68		##		
	F	8	152		#		
30	A-F	49	963	0.91			
31	A	4	83		##		
	B	7	108			*	bsk(33?), Mdh1(ca.37), J (41?/31B-E)
	C	10	142		##		sucr (ca.37/31C-F)
	D	15	249				da (41.3?31C-32A)
	E	11	205				
	F	4	132		#		abo,hup (44.0/31F-32E)
31	A-F	51	919	0.83			
32	A	6	161		#		
	B	3	78		#	*	
	C	4	108		#	**	
	D	6	171		##	*	
	E	4	98		##		
	F	5	142		##	*	
32	A-F	28	758	1.25			
33	A	7	176		#		
	B	13	254		##		esc (54.9/33B1-2)
	C	4	103				
	D	5	156				
	E	9	166				
	F	5	112				
33	A-F	43	967	1.04			
34	A	12	323		#		U2 (?/34A-B); U5 (?/34A-B)
	B	11	205		#		
	C	8	186		#		
	D	8	127			*	b (48.5?34D4-5)
	E	6	152			**	rk (48.2/34E5-35D1) j (48.7/34E5-35D);
	F	6	147		#	*	wb (?/34F)
34	A-F	51	1140	1.03			

Table 1 (continued)
COMBINED CATALOGS OF BAND NUMBERS, DNA CONTENTS, AVAILABLE CLONES AND TRANSFORMED STRAINS AS WELL AS SOME EXAMPLES OF LOCALIZED GENES IN THE SALIVARY GLAND GENOME OF *DROSOPHILA MELANOGASTER*

2L chromosome

a (D)	b (SD)	c (BN)	d (DNA-C)	e (TDF)	I (CL)	II (TR)	III (Localized genes:) (with cross over/chromosomal site)
35	A	3	103		#	*	el (50.0/35A1)
	B	12	303		##	*	Adh (50.1/35B2-3); ck (51.0/35B10-C1); tRNA Gly-3 (?/35BC)
	C	6	132		#	*	Su(H) (50.5), pu, Sco (51.0/34E5-35D1); rd (51.2/35C3-36B5)
	D	8	205			*	Epa (52/35D1-2); cru (52.5/35D5)
	E	6	93			*	
	F	12	200				
35	A-F	47	1036	1.02			
36	A	16	323			*	Ifm(2)1-3 (ca.52?/36A-B)
	B	6	156		##		Mhc 36B (ca.52?/36B)
	C	11	176		#	*	tyr 1 (52.4/36C); Bic-d (52.9/36C); M(2)m (54.0/36B6-37B2)
	D	4	103				Dox3 (53.1/36D1-E4)
	E	5	127				
	F	15	269		#		msl1 (53.3/36F7-37B8)
36	A-F	57	1154	0.94			
37	A	7	147			*	
	B	13	240		##	*	M(2)H (53.3/37B2-40B2); hk (53.9/37B10-13)
	C	7	152		##	*	Dcd, l(2)amd (53.9/37C1-2); fs(2)37C (54/37C2-D1)
	D	8	181				l(2)37BF (53.9/37D)
	E	6	166				
	F	5	112				
37	A-F	46	998	1.0			
38	A	11	225		#		U2 (?/38A-B)
	B	7	103			**	
	C	11	225			*	
	D	6	122			*	Bl (54.8/38A6-E9)
	E	9	171		#	*	
	F	7	112				
38	A-F	51	958	0.87			

Table 1 (continued)
COMBINED CATALOGS OF BAND NUMBERS, DNA CONTENTS, AVAILABLE CLONES AND TRANSFORMED STRAINS AS WELL AS SOME EXAMPLES OF LOCALIZED GENES IN THE SALIVARY GLAND GENOME OF *DROSOPHILA MELANOGASTER*

2L chromosome

a (D)	b (SD)	c (BN)	d (DNA-C)	e (TDF)	I (CL)	II (TR)	III (Localized genes:) (with cross over/chromosomal site)
39	A	7	156				
	B	3	103			**	U5 (?/39B)
	C	3	68		#	*	Tpl (ca. 54/39C-D)
	D	7	181		#		His1, His3, His4, His2a, His2, (tandem repeats) (ca.54?/39D2-F2)
	E	7	171		#	**	
	F	2	68			*	pr (54.?/37B2-40B2)
39	A-F	29	747	1.19			
40	A	7	196				Ifm(2)11(ca.55/40A)
	B	5	103				lt (55.0/40B-F); M(2)m (54?/40B3)
	C	3	64				
	E-F	?	?			*	
40	A-F	15?	363?	1.12	#		

2R chromosome

a (D)	b (SD)	c (BN)	d (DNA-C)	e (TDF)	I (CL)	II (TR)	III (Localized genes:) (with cross over/chromosomal site)
41	A	?	?				1 (2)Sp11, 1 (2) Sp15, 1 (2) Sp9, rl stw, M (2) S2 (55.1/41A); tRNATyr, Dip-A (55.2/41A-B)
	B	?	?				ap (55.2/41B-C);
	C	8	200				msf (55.2/41A-42A3)
	D	4	108				
	E	6	181				
	F	12	249				
41	A-F	30	738	1.14	#		
42	A	16	440		##	**	pk (55.3/42A3-16); tk (55.3/42A2-B9); tRNA Asn-5, Arg-2, Met-3, Tyr, Lys-2, Ile (?/42A6-16); Act42A (55.4/42A)
	B	4	122		#	*	bur (55.7/42B1-3); Hsa (?/42B)
	C	9	240		#		
	D	3	108			*	mle (55.8/41A-43A)
	E	6	156		##	**	tRNA Met-3, Tyr, Lys-2 (?/42E)
	F	3	78		#	**	
42	A-F	41	1144	1.29			ltd (56/41A-44C)

Table 1 (continued)
COMBINED CATALOGS OF BAND NUMBERS, DNA CONTENTS, AVAILABLE CLONES AND TRANSFORMED STRAINS AS WELL AS SOME EXAMPLES OF LOCALIZED GENES IN THE SALIVARY GLAND GENOME OF *DROSOPHILA MELANOGASTER*

2R chromosome

a (D)	b (SD)	c (BN)	d (DNA-C)	e (TDF)	I (CL)	II (TR)	III (Localized genes:) (with cross over/chromosomal site)
43	A	6	152		##		
	B	5	142		##		Drl (51/43B3-8)
	C	7	147		#	**	
	D	7	127		#	*	
	E	18	411		##	**	cn (57.5/43E3-14)
	F	9	171				
43	A-F	52	1150	1.02			
44	A	7	127				
	B	8	205				
	C	9	230		#	**	
	D	8	171		##	*	H44D, D44D, L44D, cp (?/44D)
	E	3	78		#	*	Pgi (58.6/44E-F?)
	F	10	196		##	*	Np (58.7/44F1-45E2)
44	A-F	45	1007	1.04	#		flz (59/44F-46D)
45	A	11	210		##	**	
	B	6	122		#	*	
	C	6	117				
	D	6	117		#	*	
	E	6	73			**	
	F	4	122				
45	A-F	36	761	0.98			
46	A	3	142				
	B	8	147		#		
	C	8	156			**	eve (59?/46C3-8)
	D	6	127		#		
	E	5	93		##		
	F	9	200		#		
46	A-F	39	865	1.03			
47	A	12	381			**	
	B	10	205				
	C	5	142			*	
	D	7	166			**	
	E	5	122		#		
	F	10	259		##		
47	A-F	49	1275	1.2			
48	A	3	142		#	*	en (62.0/48A1-4)
	B	7	122		#	**	
	C	8	142		#	*	
	D	7	132		#	*	Lcp 1-4 (ca.62/48D); Hdl-family (ca.62/48D)

Table 1 (continued)
COMBINED CATALOGS OF BAND NUMBERS, DNA CONTENTS, AVAILABLE CLONES AND TRANSFORMED STRAINS AS WELL AS SOME EXAMPLES OF LOCALIZED GENES IN THE SALIVARY GLAND GENOME OF *DROSOPHILA MELANOGASTER*

2R chromosome

a (D)	b (SD)	c (BN)	d (DNA-C)	e (TDF)	I (CL)	II (TR)	III (Localized genes:) (with cross over/chromosomal site)
	E	10	166		##	*	Deb-A, Deb-B (?/48E-F)
	F	9	147		##	*	
48	A-F	44	851	0.90			
49	A	9	225		##		Cal (67?/49A)
	B	10	225			*	Arp (67?/49A9-B3)
	C	4	112		##		
	D	5	103		##	**	sca (66.7/49D1-3); vg, bic (67.0/49D3-E6); l(2)C (67.1/49D4)
	E	6	122		##	*	
	F	11	220		##	**	
49	A-F	45	1007	1.04			cmd (?/49?)
50	A	12	220		#	*	
	B	8	122		##	**	tRNA Lys-2 (?/50B)
	C	18	416		##		tRNA Tyr-lg (?/50C)
	D	6	98		#		Gadph (?/50D-51A2)
	E	6	127				
	F	6	132		#		
50	A-F	56	1115	0.92	#		
51	A	8	191		#		Asx (72/51A1-B4)
	B	8	156		#		
	C	5	112		#		
	D	12	186		##		
	E	10	196		#		
	F	10	181				
51	A-F	53	1022	0.89			1(2)me (ca.72/51-2)
52	A	10	200		##	*	Hex-C (73.5/51-2?)
	B	5	93		#	**	
	C	9	166			**	
	D	14	191		##	*	Gpo (ca. 75/52D); M(2)S7 (77.5/52D)
	E	9	127		#		Got1 (75.0/52D-E?)
52	A-F	59	977	0.77			
53	A	5	117			*	
	B	6	103			*	
	C	15	381		##	*	
	D	15	264		##		
	E	9	191			**	
	F	14	230		#	**	
53	A-F	64	1286	0.93	#		

Table 1 (continued)
COMBINED CATALOGS OF BAND NUMBERS, DNA CONTENTS, AVAILABLE CLONES AND TRANSFORMED STRAINS AS WELL AS SOME EXAMPLES OF LOCALIZED GENES IN THE SALIVARY GLAND GENOME OF *DROSOPHILA MELANOGASTER*

2R chromosome

a (D)	b (SD)	c (BN)	d (DNA-C)	e (TDF)	I (CL)	II (TR)	III (Localized genes:) (with cross over/chromosomal site)
54	A	3	68		#	**	Amy-p, Amy-d (77.7/54A1-B1)
	B	17	318		#		
	C	10	205			*	
	D	6	132				
	E	10	186		#		Hsp54E (?/54E1)
	F	5	88		##		map (80/?); Phox (80.6/?)
54	A-F	51	997	0.91			
55	A	4	196		#		
	B	10	215		#		Eip40(Eip55BD)(?/55B-D)
	C	11	284		#		
	D	3	103		#		Aldox2 (ca.86/?)
	E	12	298				tRNA Gly3 (?/55E)
	F	11	220		#		sdh (ca. 89/?)
55	A-F	51	1316	1.19			
56	A	3	93			*	
	B	6	171			*	
	C	9	215		#		β Tub 56C (?/56C)
	D	18	342		##	*	β Tub 56D (?/56D4-12)
	E	6	147		#		hy, tRNA Glu4, His (?/56F)
	F	19	552		##	**	5SRNA genes (ca. 93/56F); tRNA Gly3, Lys2, Met3, Phe2, Thr6, Tyr (?/56EF)
56	A-F	61	1520	1.15			
57	A	11	274			**	Act 57A (93/57A)
	B	19	513		#	**	hy (93.3/57A-F)
	C	9	161		##		Pu (ca.97/57C1-8)
	D	13	240				Elp (99/57C9-D5)
	E	8	147				
	F	13	220		#	*	
57	A-F	73	1555	0.99			
58	A	3	117				l(2)Su(H) (99.0/58A1)
	B	8	171				
	C	6	142		#		
	D	8	220			**	Adk-C (ca. 100/?)
	E	10	235			*	crs (?/58E3-59A2); px (100.5/58E-F)
	F	6	166		#	**	pa (101.0/58F2); M(2)I (101.2/58F)
58	A-F	41	1051	1.19			inaD (101/58F-60F1)

Table 1 (continued)
COMBINED CATALOGS OF BAND NUMBERS, DNA CONTENTS, AVAILABLE CLONES AND TRANSFORMED STRAINS AS WELL AS SOME EXAMPLES OF LOCALIZED GENES IN THE SALIVARY GLAND GENOME OF *DROSOPHILA MELANOGASTER*

2R chromosome

a (D)	b (SD)	c (BN)	d (DNA-C)	e (TDF)	I (CL)	II (TR)	III (Localized genes:) (with cross over/chromosomal site)
59	A	4	142				
	B	9	230			*	
	C	7	176			*	l(2)bw (104/59CD)
	D	11	298			*	bw (104.5/59D4-E1)
	E	5	147		#	*	mi (104.7/59E1-2); abb (105.5/59E2); pd (106.4/59E2-60B10)
	F	9	249				ll, mr (106.7/59E2-60B10)
59	A-F	45	1242	1.28			
60	A	17	347		##	**	
	B	15	323		##	**	l(2)ax (106.9/60B); Dat (107/60B1-10); E(w[a]) (107?/60B10); l(2)NS (107.0/60B10-12)
	C	10	225		##	**	Fo (107.0/58E-60B10); Sp (107.0/60B13-C5); tRNA Asn-5 (?/60C)
	D	16	381				Px (107.2/60C6-D1); bs (107.3/60C5-D2); Pin (107.3/60C5-D2); ba (107.4/60C5-D2)
	E	11	332			**	Ba (107.8/60D-E); M(2)c (108?/60E3-11); gsb (107.6?/60E9-F1); Kpn (107.6?/60F2-5); If (107.6?/60F3)
	F	5	210		##	*	
60	A-F	74	1818	1.14			

3L chromosome

a (D)	b (SD)	c (BN)	d (DNA-C)	e (TDF)	I (CL)	II (TR)	III (Localized genes:) (with cross over/chromosomal site)
61	A	8	191		##	**	LspIg (-1.4/61A1-6)
	B	8	142				su(ve) (?/61A-E)
	C	17	372			*	
	D	5	103			**	
	E	4	103			*	aa (0.0?/61E2-62A6)
	F	15	230		#	*	ru (0.0?/61F5-62A3); ve (0.2/61E2-62A6); tRNA Thr6 (?/61F)
61	A-F	57	1141	0.93			
62	A	17	318		##	**	tRNA lys2, Gly4 (?/62A)
	B	17	313		##	*	Aprt (3.0/62B7-12); C-ras3 (15/34B)

Table 1 (continued)
COMBINED CATALOGS OF BAND NUMBERS, DNA CONTENTS, AVAILABLE CLONES AND TRANSFORMED STRAINS AS WELL AS SOME EXAMPLES OF LOCALIZED GENES IN THE SALIVARY GLAND GENOME OF *DROSOPHILA MELANOGASTER*

3L chromosome

a (D)	b (SD)	c (BN)	d (DNA-C)	e (TDF)	I (CL)	II (TR)	III (Localized genes:) (with cross over/chromosomal site)
	C	6	166		#		
	D	12	205		##		dib (12?/62D-64C?)
	E	12	249		#		
	F	6	142				
62	A-F	70	1393	0.92			
63	A	11	230		#	*	U5 (?/63A)
	B	14	235		##	**	Hsp83 (?/62B-C1)
	C	8	152		##	**	
	D	5	122				
	E	10	230		#	*	
	F	6	137		##		Hsp63F (?/63F1)
63	A-F	54	1106	0.95	#		
64	A	15	298		#		
	B	18	303		##	*	C-ras1, C-ras2 (15/64B); Ama1 (19/64B-C)
	C	16	332		##	**	jv, Me(19.2), dv(20.0), me (29.0?/64C12-65E1)
	D	8	152			*	tRNA Ser4, Ser7, Val3a (?/64D)
	E	10	156				
	F	5	127		##	**	
64	A-F	72	1368	0.88	#		
65	A	15	328		#	**	
	B	8	210		#	*	
	C	4	117		#		
	D	7	156			**	
	E	14	308				
	F	12	181			**	
65	A-F	60	1300	1.0	#		
66	A	28	518			**	Hn (23.0/66A-B)
	B	17	225		#		Argk (25.2/66B-D11)
	C	11	230		##		Idh (25.4/?)
	D	21	372		##	*	msl 3 (26.0/66D?); h (26.5/66D2-E1);
	E	10	172			**	Cp15, 16, 18, 19 (=s15, s16, s18, s19) (26.5/66D11-15)
	F	7	127		#	*	
66	A-F	94	1644	0.81	#		

Table 1 (continued)
COMBINED CATALOGS OF BAND NUMBERS, DNA CONTENTS, AVAILABLE CLONES AND TRANSFORMED STRAINS AS WELL AS SOME EXAMPLES OF LOCALIZED GENES IN THE SALIVARY GLAND GENOME OF *DROSOPHILA MELANOGASTER*

3L chromosome

a (D)	b (SD)	c (BN)	d (DNA-C)	e (TDF)	I (CL)	II (TR)	III (Localized genes:) (with cross over/chromosomal site)
67	A	12	210		#	*	Difl (?/67A8-10)
	B	20	254		##	*	Hsp22, 23, 26, 28 (?/67B); Idh 1 (27.1/66B-67C)
	C	17	298		##	*	αTub 67C (?/67C4-6)
	D	15	259		#		
	E	11	196		#	*	
	F	4	108		#		lxd (34.5/?); sod (34.6/?)
67	A-F	79	1325	0.78			
68	A	13	240			*	rs (35.0/?)
	B	4	68				αFuc (35.5/?)
	C	20	337		##	**	Sgs3, 7, 8 (ca. 36/68C3-5); rt (?/68C9-12)
	D	9	152			*	
	E	5	137		##	*	Lsp2 (36 + ?/68E3-4)
	F	10	215		#		
68	A-F	61	1149	0.87			
69	A	6	142				Est6 (36.8/69A1-5); app (37.5/69A2-5)
	B	7	142				
	C	10	166			*	gsp, eyg (?/69C1-5)
	D	8	156		#	*	
	E	8	147				
	F	9	147		#		
69	A-F	48	900	0.87	#		
70	A	8	210		##	*	M(3)h (40.2/70A); Ly (40.5/70 A3-5); tRNA Asp2 (?/70A)
	B	7	166		##	*	D (40.7/69D3-70D1)
	C	15	308		##	*	Hsc1 (?/70C); Gl (41.4/70C2) tRNA Val3a Val4 (?/70C)
	D	9	235		##		P1, P6 (?/70C-D)
	E	9	176				
	F	9	166				tRNA Met3 (?/70F)
70	A-F	57	1261	1.02			
71	A	4	141.8		#	*	
	B	13	224.9		#	*	
	C	7	166.2		##	*	Eip28,29 (?/71C3-D2)
	D	6	88.0		##		Sgs6 (42.0/71C-F5)
	E	8	171.1		##		
	F	8	132.0			**	

Table 1 (continued)
COMBINED CATALOGS OF BAND NUMBERS, DNA CONTENTS, AVAILABLE CLONES AND TRANSFORMED STRAINS AS WELL AS SOME EXAMPLES OF LOCALIZED GENES IN THE SALIVARY GLAND GENOME OF *DROSOPHILA MELANOGASTER*

3L chromosome

a (D)	b (SD)	c (BN)	d (DNA-C)	e (TDF)	I (CL)	II (TR)	III (Localized genes:) (with cross over/chromosomal site)
71	A-F	46	924.0	0.93			
72	A	6	146.7				th (43.2/72A2-E5)
	B	2	53.8		#		
	C	3	78.2		#		
	D	13	308.0		#		Pgm (43.4/72D1-5)
	E	5	146.7		#		bud (?/ 72E-73A)
	F	3	73.3				tRNA Met2 (?/72F1-2)
72	A-F	32	806.7	1.17			
73	A	13	293.3		#		tRNA Met2 (?/73A1-2); st (?/73A2-B1)
	B	8	156.4		#		Dash (ca. 44/73B)
	C	5	88.0				
	D	8	122.2		#		Hsp 73D (?/73D1)
	E	7	156.4		#		
	F	7	102.7		#		
73	A-F	48	919.0	0.89			
74	A	6	166.2				Ars (?/74A-79D)
	B	5	97.8				
	C	3	102.7				
	D	5	127.1				
	E	5	117.3		#		Ei RNA (?/71E-F)
	F	4	83.1		#		
74	A-F	28	694.2	1.15			
75	A	10	273.8				
	B	12	234.6				
	C	10	195.5		#	*	
	D	8	122.2			**	Cat (44.3/75D-76A)
	E	7	132.0				
	F	11	180.9				
75	A-F	58	1139.0	0.91			
76	A	6	161.3		#	*	
	B	9	259.1				
	C	7	156.4				
	D	9	224.9		#		
	E	4	127.1		#		
	F	3	112.4		#		
76	A-F	38	1041.2	1.27			DNase (45.9/76-77)

Table 1 (continued)

COMBINED CATALOGS OF BAND NUMBERS, DNA CONTENTS, AVAILABLE CLONES AND TRANSFORMED STRAINS AS WELL AS SOME EXAMPLES OF LOCALIZED GENES IN THE SALIVARY GLAND GENOME OF *DROSOPHILA MELANOGASTER*

3L chromosome

a (D)	b (SD)	c (BN)	d (DNA-C)	e (TDF)	I (CL)	II (TR)	III (Localized genes:) (with cross over/chromosomal site)
77	A	5	137			*	
	B	11	220				in (46-47?/77B-C)
	C	7	156				
	D	4	78			*	
	E	9	215			*	kni (46?/77E)
	F	5	103				
77	A-F	41	909	1.03			
78	A	8	210				
	B	4	103			*	
	C	11	249			**	
	D	10	205		#	**	
	E	6	132				
	F	4	88				
78	A-F	43	987	1.06			Aph1 (47.3/78-79?)
79	A	7	142				eg (47.3/79A4-B1)
	B	4	93		#		Act 79B (ca. 47.5/79B)
	C	3	73				
	D	4	103				
	E	8	235		#	*	
	F	8	210			*	tRNA Leu2 (?/79F)
79	A-F	34	856	1.16	#		
80	A	4	112			*	
	B	3	78				
	C	5	132		#		
	D	?	?				
	E	?	?				
	F	?	?		#		
80	A-C	12	322	1.24	#		

3R chromosome

a (D)	b (SD)	c (BN)	d (DNA-C)	e (TDF)	I (CL)	II (TR)	III (Localized genes:) (with cross over/chromosomal site)
81	F	6	191		#		
82	A	8	176		#	*	
	B	5	73			**	
	C	7	127			**	
	D	10	156				
	E	9	191		##		U1 (?/82E)
	F	10	186		#	**	

Table 1 (continued)
COMBINED CATALOGS OF BAND NUMBERS, DNA CONTENTS, AVAILABLE CLONES AND TRANSFORMED STRAINS AS WELL AS SOME EXAMPLES OF LOCALIZED GENES IN THE SALIVARY GLAND GENOME OF *DROSOPHILA MELANOGASTER*

3R chromosome

a (D)	b (SD)	c (BN)	d (DNA-C)	e (TDF)	I (CL)	II (TR)	III (Localized genes:) (with cross over/chromosomal site)
82	A-F	49	909	0.86			
83	A	8	181		##	*	
	B	9	191		##	**	
	C	8	176		##	*	
	D	6	176		#		Ki (47.6/83DE)
	E	8	191				
	F	3	83		#	**	tRNA Met3 (?/83F3-4)
83	A-F	42	998	1.10			
84	A	9	269		##		Dfd (47.5?/84A-B)
	B	7	171		##	**	Antp, Hu, ftz (47.8/84B1-2); αTub 84B (47.8/84B3-6)
	C	8	156		##	**	Gld (48/84C-D);
	D	12	298		##	**	tRNA Val3, Val3b (?/84D); EstC (48/84D3-12); aTub 84D (ca.48/84D4-8); Dipr (?/84D-F)
	E	11	186		##	**	dsx (ca. 48/84E1-2)
	F	14	215		##	**	tRNA Asn5, Arg2 (?/84F)
84	A-F	61	1295	0.98			
85	A	10	279		##	**	p (48.0/84A6-B3); hb (?/85A)
	B	9	235			**	Dhod (ca. 48+/ 84A-C)
	C	14	230		##		Ali (48.3/?)
	D	25	430		##	**	B2t (48.5/85D7-11); by (48.7/85D11-E13); C-ras1 (49/85D)
	E	14	318		##		Mtn (ca. 49/85E; hth (48/85E-86B) α Tub 85E (ca. 49/85E6-10)
	F	17	352			**	knk (49/85E-86B)
85	A-F	89	1844	0.96	#		
86	A	8	156			*	tRNA Ser2b (?/86A)
	B	5	93		#		
	C	17	284			*	
	D	10	196			**	Odh (49.2/86D1-4)
	E	21	411			*	cu (50.0/86D2-87B2)
	F	10	181				
86	A-F	71	1321	0.86	#		
87	A	12	240		##	**	Hsp70 (ca. 51/87A7,2 genes)
	B	22	372		#	**	Dip-C (ca. 51/87B5-10); tRNA Lys2, Lys5 (?/87B)

Table 1 (continued)
COMBINED CATALOGS OF BAND NUMBERS, DNA CONTENTS, AVAILABLE CLONES AND TRANSFORMED STRAINS AS WELL AS SOME EXAMPLES OF LOCALIZED GENES IN THE SALIVARY GLAND GENOME OF *DROSOPHILA MELANOGASTER*

3R chromosome

a (D)	b (SD)	c (BN)	d (DNA-C)	e (TDF)	I (CL)	II (TR)	III (Localized genes:) (with cross over/chromosomal site)
	C	12	215		##	*	Hsp70 (3 genes), Hsr aB (ca. 51/87C1); kar (51.7/87C1); tRNA Thr3 (?/87C)
	D	16	342		##	*	Men (51.7/87D1-2); Hsc2 (?/87D); ry (52.0/87D1-2); snk (52.0/87D10-12)
	E	12	249		##	**	Ace (52.2/87E1-2); Act87E, Mhc87E (52.3/87E)
	F	15	274		#	**	Dip-B (53.6/87F1-88C3); atn (54.0/87F2-?)
87	A-F	89	1692	0.88			
88	A	11	230			*	ems (53/88A1-10); red (53.6/88A-C)
	B	8	142		#		Hsp 88B (?/88B1); cv-c (54.1/88A-C); tRNA Ser 2b (?/88B)
	C	12	240		#	*	
	D	12	235		#		Ifm (3)1-7 (ca. 55/?)
	E	15	298		#	**	hsc3 (?/88E); R3-55.4 (55.4/?)
	F	8	186		##	**	m-Est (56.7/?); Act88F, Mhc 88F (ca. 57/88F)
88	A-F	66	1331	0.93	#		
89	A	12	220			**	lpo, Aldox1 (57.2/89A)
	B	24	460		##	**	c(3)G (57.4/88F9-89B5); Sb, sbd (58.2/89B4-5)
	C	8	196		#		ss (58.5/89C1-2)
	D	9	196				
	E	15	298		##		bx, Cbx, Ubx, bxd, pbx (58.8/89E1-4); Mc (59.0/89E7-11)
	F	4	88				
89	A-F	72	1458	0.94			
90	A	6	117				
	B	8	166		##		tRNA Val3b, Ala (?/90B-C)
	C	12	200		##	*	DNASE1 (61.8/90C2-E); fru (62/90C-91A)
	D	4	73			*	sr (62.0/90D2-F7)
	E	7	112			**	
	F	11	147			*	

Table 1 (continued)
COMBINED CATALOGS OF BAND NUMBERS, DNA CONTENTS, AVAILABLE CLONES AND TRANSFORMED STRAINS AS WELL AS SOME EXAMPLES OF LOCALIZED GENES IN THE SALIVARY GLAND GENOME OF *DROSOPHILA MELANOGASTER*

3R chromosome

a (D)	b (SD)	c (BN)	d (DNA-C)	e (TDF)	I (CL)	II (TR)	III (Localized genes:) (with cross over/chromosomal site)
90	A-F	48	815	0.79			
91	A	7	152				
	B	9	235			*	Kf1 (?/91B-93F)
	C	6	117		#	**	Cha (64.6/91C)
	D	6	156		#	*	Dl (66.2/91D-92A2)
	E	5	122				
	F	12	220			*	
91	A-F	45	1002	1.03	#		
92	A	14	259		##	**	ninaE (ca. 66-67/92A-B)
	B	11	200		##	**	tRNA Val3b (?/92B1-2)
	C	6	137		#	*	
	D	8	137		#		H (69.5/92D-94A)
	E	12	220		#		
	F	10	230			*	
92	A-F	61	1183	0.90	#		
93	A	6	127			*	
	B	12	225		#	*	r-l (ca. 70/93B4-C)
	C	7	127		#		e (70.7/93B7-F9)
	D	10	186		#	**	Hsr 93D (?/93D4-9)
	E	11	210				
	F	14	386				
93	A-F	60	1261	0.97			
94	A	18	318		#		tRNA Ser2b (?/94A)
	B	10	196			*	cd (75.7/94A-E)
	C	7	147				
	D	12	269		#	*	bar-3 (79.1/94A-E)
	E	10	244		#	*	hh (81/94E)
	F	5	132		#		
94	A-F	62	1306	0.97			
95	A	7	176		#	**	
	B	7	137		#	*	
	C	13	249			*	U1 (?/95C)
	D	8	225		##	**	Hsp 68 (?/95D)
	E	7	171				crb (82/95E-96A)
	F	14	259			**	
95	A-F	56	1217	1.01			Gdh (81.7/95?)
96	A	26	440		#	*	tRNA Ser2b (?/96A)
	B	9	127		##	**	

Table 1 (continued)
COMBINED CATALOGS OF BAND NUMBERS, DNA CONTENTS, AVAILABLE CLONES AND TRANSFORMED STRAINS AS WELL AS SOME EXAMPLES OF LOCALIZED GENES IN THE SALIVARY GLAND GENOME OF *DROSOPHILA MELANOGASTER*

3R chromosome

a (D)	b (SD)	c (BN)	d (DNA-C)	e (TDF)	I (CL)	II (TR)	III (Localized genes:) (with cross over/chromosomal site)
	C	11	181		##		
	D	6	98		#		
	E	12	240				E(spl) (89/96E-F)
	F	12	357		##		
96	A-F	76	1443	0.88	#	*	
97	A	8	225		##	**	Ald (91.5/97A-B)
	B	8	166		#	**	
	C	5	137		#		rsd (95.4/?)
	D	13	284		#		
	E	10	191		#		Bd (93.8/97E-F); rsd (95.4/?)
	F	8	152		##*	βTub 97F (?/ 97F)	
97	A-F	52	1155	1.03			
98	A	11	205			*	
	B	6	161			*	fkh (98/98B-99A)
	C	5	127			**	
	D	6	181				
	E	7	156		#		
	F	16	264		##		
98	A-F	51	1094	0.99	#		Lap-A, B, C, D, (98.3/98)
99	A	10	279			**	kay (99/99A-100A); Ama2 (100/?)
	B	11	240		#	*	
	C	8	142		##		ca. (100.7/99C5-9)
	D	8	156		##	**	Acph1 (101.1/99B5-8); Rbp49 (?/99D)
	E	7	152		##	*	Tpi (101.3/99B-E); Mlc 99E (?/99E)
	F	11	230		#		
99	A-F	55	1199	1.01	#		BGlu (101?/98F-100F)
100	A	8	235		#		
	B	11	352		##		
	C	7	225		##		
	D	6	127		##	*	
	E	3	68		#		
	F	7	202		##	*	
100	A-F	42	1209	1.33	#		

Table 1 (continued)
COMBINED CATALOGS OF BAND NUMBERS, DNA CONTENTS, AVAILABLE CLONES AND TRANSFORMED STRAINS AS WELL AS SOME EXAMPLES OF LOCALIZED GENES IN THE SALIVARY GLAND GENOME OF *DROSOPHILA MELANOGASTER*

3R chromosome

a (D)	b (SD)	c (BN)	d (DNA-C)	e (TDF)	I (CL)	II (TR)	III (Localized genes:) (with cross over/chromosomal site)
			4. Chromosome				
101	C-F	10	259				ar (?/101E-102B); Ce, Scn, l(4)1 (?/101E-102B6); l(4)13, ci, l(4)18, l(4)25, M(4) (?/101F-102A5)
102	A	6	161				bt, gy (?/102B10-E9)
	B	7	274				
	C	9	225		##		
	D	6	230		##		spa (?/102D-F)
	E	5	161		#		sv (3.0/102E2-F7); ey, fs(4)34 (2.0/102E2-F10)
	F	7	235		##		
102	A-F	40	1286	1.49	#		

Note: Band numbers per subdivisions were counted according to the interpretation given in the EM maps. DNA contents per single chromatids of salivary gland chromosomes were calculated by using the same thickness values of bands as were used in drawing the EM maps. The tentative density correction coefficients (called as tentative density factors ≫ = TDF) were calculated for the DNA values of the divisions by using the relative thickness of bands as an indicator of relative density of the band chromatin. Explanations of the Columns I to III. (I) Cloned DNA, (# = cloned DNA available, ## = several clones); (II) transformed strains, (* = available, ** = several strains). For more detailed information of clones and inserts, see the original catalogs of J. Merriam et al.[52,53,56] published in *DIS (Drosophila Information Service)*. The data for Column III (localized genes) has been collected from publications[28,54,55] and completed with data from a recent computer list of gene sites kindly provided by Dr. D. L. Lindsley. Explanation of the other columns: a = division number, b = subdivision, c = number of bands according to the EM map, d = estimated DNA content per chromatid, e = tentative correction coefficient for DNA contents of divisions (can be calculated also for subdivisions by using the Formula 3 given above).

REFERENCES

1. **Beermann, W. and Bahr, G.,** The submicroscopic structure of Balbiani ring, *Exp. Cell Res.*, 6, 195, 1954.
2. **Gay, H.,** Serial sections of smears for electron microscopy, *Stain Technol.*, 30, 239, 1955.
3. **Swift, H.,** Nucleic acids and cell morphology in Dipteran salivary glands, in *The Molecular Control of Cellular Activity*, Allen, M. M., Ed., MacGraw-Hill, New York, 1962, 73.
4. **Sorsa, M. and Sorsa, V.,** The squash technique in the electron microscopic studies on the structure of polytene chromosomes, *J. Ultrastruct. Res.*, 20, 302, 1967.
5. **Sorsa, M. and Sorsa, V.,** Electron microscopic observations on interband fibrils in *Drosophila* salivary chromosomes, *Chromosoma*, 22, 32, 1967.
6. **Berendes, H. D.,** Electron microscopical mapping of the giant chromosomes, *Drosophila Inf. Serv.*, 43, 115, 1968.

7. **Sorsa, M.,** Ultrastructure of the chromocentre heterochromatin in *Drosophila melanogaster, Ann. Acad. Sci. Fenn., A IV. Biol.,* 146, 1, 1969.
8. **Sorsa, M.,** Ultrastructure of puffs in the proximal part of chromosome 3R in *Drosophila melanogaster, Ann. Acad. Sci. Fenn., A IV, Biol.,* 150, 1, 1969.
9. **Sorsa, M.,** Ultrastructure of the polytene chromosome in *Drosophila melanogaster,* with special reference to electron microscopic mapping of chromosome 3R, *Ann. Acad. Sci. Fenn., A IV, Biol.,* 151, 1969.
10. **Berendes, H. D.,** Polytene chromosome structure at submicroscopic level. I. A map of region X, 1-4E of *Drosophila melanogaster, Chromosoma,* 29, 118, 1970.
11. **Sorsa, V., Green, M. M., and Beermann, W.,** Cytogenetic fine structure and chromosomal localization of the white gene in *Drosophila melanogaster, Nature New Biol.,* 245, 43, 1973.
12. **Sorsa, V.,** Ultrastructure of the 5S RNA locus in the salivary gland chromosomes of *Drosophila melanogaster, Hereditas,* 74, 297, 1973.
13. **Zhimulev, I. F., Semeshin, V. F., and Belyaeva, E. S.,** Fine cytogenetical analysis of the band 10A1-2 and adjoining regions in the *Drosophila melanogaster* X chromosome. I. Cytology of the region and mapping of chromosome rearrangements, *Chromosoma,* 82, 9, 1981.
14. **Semeshin, V. F. and Szidonya, J.,** EM mapping of rearrangements in the 24-25 sections of *Drosophila melanogaster* 2L chromosomes, *Drosophila Inf. Serv.,* 61, 148, 1985.
15. **Semeshin, V. F., Baricheva, E. M., Belyaeva, E. S., and Zhimulev, I. F.,** Electron microscopical analysis of *Drosophila* polytene chromosomes. I. Mapping of the 87A and 87C heat shock puffs in development, *Chromosoma,* 87, 229, 1982.
16. **Semeshin, V. F., Baricheva, E. M., Belyaeva, E. S., and Zhimulev, I. F.,** Electron microscopical analysis of *Drosophila* polytene chromosomes. II. Development of complex puffs, *Chromosoma,* 91, 210, 1985.
17. **Semeshin, V. F., Baricheva, E. M., Belyaeva, E. S., and Zhimulev, I. F.,** Electron microscopical analysis of *Drosophila* polytene chromosomes. III. Mapping of puffs developing from one band, *Chromosoma,* 91, 234, 1985.
18. **Saura, A. O. and Sorsa, V.,** Electron microscopic analysis of the banding pattern in the salivary gland chromosomes of *Drosophila melanogaster:* Divisions 21 and 22 of 2L, *Hereditas,* 90, 39, 1979.
19. **Saura, A. O. and Sorsa, V.,** Electron microscopic analysis of the banding pattern in the salivary gland chromosomes of *Drosophila melanogaster:* Divisions 27, 28 and 29 of 2L, *Hereditas,* 91, 219, 1979.
20. **Saura, A. O. and Sorsa, V.,** Electron microscopic analysis of the banding pattern in the salivary gland chromosomes of *Drosophila melanogaster:* Divisions 30 and 31 of 2L, *Hereditas,* 90, 257, 1979.
21. **Saura, A. O. and Sorsa, V.,** Electron microscopic analysis of the banding pattern in the salivary gland chromosomes of *Drosophila melanogaster:* Divisions 37, 38, and 39 of 2L, *Hereditas,* 91, 5, 1979.
22. **Sorsa, V. and Saura, A. O.,** Electron microscopic analysis of the banding pattern in the salivary gland chromosomes of *Drosophila melanogaster:* Divisions 1 and 2 of X, *Hereditas,* 92, 73, 1980.
23. **Sorsa, V. and Saura, A. O.,** Electron microscopic analysis of the banding pattern in the salivary gland chromosomes of *Drosophila melanogaster:* Divisions 3, 4 and 5 of X, *Hereditas,* 92, 341, 1980.
24. **Saura, A. O.,** Electron microscopic analysis of the banding pattern in the salivary gland chromosomes of *Drosophila melanogaster:* Divisions 23 through 26 of 2L, *Hereditas,* 93, 295, 1980.
25. **Saura, A. O.,** Electron microscopic analysis of the banding pattern in the salivary gland chromosomes of *Drosophila melanogaster:* Divisions 32 through 36 of 2L, *Hereditas,* 99, 89, 1983.
26. **Sorsa, V., Saura, A. O., and Heino, T. I.,** Electron microscopic analysis of the banding pattern in the salivary gland chromosomes of *Drosophila melanogaster.* Divisions 6 through 10 of X, *Hereditas,* 98, 181, 1983.
27. **Sorsa, V., Saura, A. O., and Heino, T. I.,** Electron microscopic map of divisions 61, 62, and 63 of the salivary gland 3L chromosome in *Drosophila melanogaster, Chromosoma,* 90, 177, 1984.
28. **Lossinsky, A. S. and Lefever, H. M.,** Ultrastructural banding observations in region 1A-10F of the salivary gland X chromosome of *Drosophila melanogaster, Drosophila Inf. Serv.,* 53, 126, 1978.
29. **Grond, C. J. and Derksen, J.,** The banding pattern of the salivary gland chromosomes of *Drosophila hydei, Eur. J. Cell Biol.,* 30, 144, 1983.
30. **Kerkis, A. Yu., Zhimulev, I. f., and Belyaeva, E. S.,** EM autoradiographic study of ^{3}H-uridine incorporation into *D. melanogaster* salivary gland chromosomes, *Drosophila Inf. Serv.,* 52, 14, 1977.
31. **Semeshin, V. F., Zhimulev, I. F., and Belyaeva, E. S.,** Electron microscopic autoradiographic study on transcriptional activity of *Drosophila melanogaster* polytene chromosomes, *Chromosoma,* 73, 163, 1979.
32. **Sorsa, V.,** An attempt to estimate DNA content and distribution in the *zeste-white* region of the X chromosome of *Drosophila melanogaster, Biol. Zbl.,* 101, 81, 1982.
33. **Sorsa, V.,** Volume of chromatin fibers in interbands and bands of polytene chromosomes, *Hereditas,* 97, 103, 1982.
34. **Ris, H. and Korenberg, J.,** Chromosome structure and levels of chromosome organization, in *Cell Biology,* Vol. 2, Prescott, D. and Goldstein, L., Eds., Academic Press, New York, 1979, 268.

35. **Manning, J. E., Hershey, N. D., Broker, T. R., Pellegrini, M., Mitchell, H. K., and Davidson, N.,** A new method of in situ hybridization, *Chromosoma,* 53, 107, 1975.
36. **Wu, M. and Davidson, N.,** A transmission electron microscopic method for gene mapping on polytene chromosomes by in situ hybridization, *Proc. Natl. Acad. Sci. U.S.A.,* 78, 7059, 1981.
37. **Wu, M. and Waddel, J.,** Transmission electron microscopic study of polytene chromosome 2R from *Drosophila melanogaster, Chromosoma,* 86, 299, 1982.
38. **Ananiev, E. V., Barsky, V. E., Ilyin, Yu. V., and Ryzic, M. V.,** The arrangement of transposable elements in the polytene chromosomes of *Drosophila melanogaster, Chromosoma,* 90, 366, 1984.
39. **Ananiev, E. V. and Barsky, V. E.,** *Electron Microscopic Map of the Polytene Chromosomes in Salivary Glands of Drosophila, (D. melanogaster),* Nauka, Moscow, 1984 (in Russian).
40. **Burkholder, G. D.,** Whole mount electron microscopy of polytene chromosomes from *Drosophila melanogaster, Can. J. Genet. Cytol.,* 18, 67, 1976.
41. **Alanen, M. and Sorsa, V.,** Identifiable regions of *Drosophila melanogaster* polytene chromosomes visualized by whole mount electron microscopy, *Hereditas,* 89, 257, 1978.
42. **Alanen, M.,** Whole mount electron microscopic maps of *Drosophila melanogaster* polytene chromosomes. I. The salivary gland X-chromosome, *Hereditas,* 104, 91, 1986.
43. **Alanen, M. and Saura, A. O.,** Whole mount electron microscopy of *Drosophila melanogaster* salivary gland chromosome 2L: Divisions 21-24, *Hereditas,* 105, 41, 1986.
44. **Kalisch, W.-E. and Hägele, K.,** Surface spreading of polytene chromosomes, *Eur. J. Cell Biol.,* 23, 317, 1981.
45. **Kalisch, W.-E. and Hägele, K.,** A new spreading technique for polytene chromosomes and its efficiency for autoradiography including in situ hybridization, in *Advances in Genetics, Development and Evolution of Drosophila,* Lakovaara, S., Ed., Plenum, New York, 1982, 1.
46. **Kalisch, W.-E.,** EM chromosome mapping using surface-spread polytene chromosomes, *Genetica,* 60, 21, 1982.
47. **Kalisch, W.-E., Whitmore, Th., and Schwitalla, G.,** Electron microscopic map of surface-spread polytene chromosomes in *Drosophila hydei, Chromosoma,* 92, 265, 1985.
48. **Ananiev, E. V. and Barsky, V. E.,** Elementary structures in polytene chromosomes of *Drosophila melanogaster, Chromosoma,* 93, 104, 1985.
49. **Reiling, H., Kalisch, W.-E., Whitmore, Th., and Tegtmeier, K.,** Computer map of salivary gland chromosomes in *Drosophila hydei, Eur. J. Cell Biol.,* 34, 336, 1984.
50. **Reiling, H., Kalisch, W.-E., and Whitmore, Th.,** Computerized EM chromosome maps, *Drosophila Inf. Serv.,* 60, 172, 1984.
51. **Whitmore, Th., Kalisch, W.-E. and Reiling, H.,** An EM map of chromosome 6 of *D. hydei, Drosophila Inf. Serv.,* 60, 208, 1984.
52. **Merriam, J.,** Cloned DNA by chromosome location, *Drosophila Inf. Serv.,* 61, 9, 1985.
53. **Merriam, J.,** Transformed lines available for cloning or deleting DNA at specific chromosome sites, *Drosophila Inf. Serv.,* 61, 19, 1985.
54. **Treat-Clemons, L. G. and Doane, W. W.,** Biochemical loci of the "fruit fly" *(Drosophila melanogaster), Drosophila Inf. Serv.,* 60, 17, 1984.
55. **Merriam, J., Smalley, S. L., Merriam, A., and Dawson, B.,** The molecular genome of *Drosophila melanogaster,* catalogs of cloned DNA, breakpoints and transformed inserts by chromosome location, *Drosophila Inf. Serv.,* 63, 173, 1986.
56. **Merriam, J., Smalley, S. L., Merriam, A., and Dawson, B.,** The molecular genome of *Drosophila melanogaster,* catalogs of cloned DNA, breakpoints and transformed inserts by chromosome location, *Drosophila Inf. Serv.,* 63, 173, 1986.

Chapter 17

CHROMOSOMAL MAPPING OF GENES IN *DROSOPHILA MELANOGASTER*

The localization of transcriptional activity in polytene chromosomes has been already discussed in some extent in Chapters 11 and 12 in Volume I. In the following chapter, the problems concerning the number and localization of genes are discussed further in light of newest results obtained mainly by using the molecular methods.

I. THE PROPORTION OF "GENE DNA" AND THE NUMBER OF GENES

The amount of potential "gene DNA" including both the single copies (unique sequences) and the gene families presumably corresponds to the proportions of SC (single copy) DNA and MR (moderately repetitive) DNA in the genome of *Drosophila melanogaster*. According to the diagram shown in Volume I, Chapter 7, Figure 2, these DNA classes (SC and MR) contain roughly about 80% of the total DNA of *D. melanogaster* genome. However, as shown by the localization and analysis of the size of genes in the heavy bands 3C2-3 and 10A1-2 the proportion of gene DNA is much lower than the total DNA content of chromomeres forming these bands.[1-5]

The DNA affecting on the only recognized genetic function of the double band 3C2-3 is obviously included in the *white*-locus, and it is apparently less than 5% of the total DNA content of the chromomeres 3C2-3 in single chromatids. Accordingly, the proportion of DNA included in the three or four genes localized in the area of the complex band 10A1-2, is probably less than 20% of the estimated total DNA content of chromomeres forming this "doublet" 10A1-2 in the salivary gland chromosomes of *D. melanogaster*.

Results from the mutation saturation experiments on *D. melanogaster*[6-8] have strongly supported the principle of equal number of genes and bands (or interbands) in any given region of polytene chromosomes. This principle suggests that the total number of genes is roughly 5000 in the genome of *D. melanogaster*. This number certainly excludes the extra copies of repetitive genes forming the gene families. The number (about 5000) can be considered as the minimum number of different genes in the genome of *Drosophila*.

Presuming that the DNA length of those about 5000 "chromomeric" genes varies from 1 to 5 kb (including possible introns), having an average length of about 3 kb, the proportion of DNA included in these genes is about 15,000 kb. This is roughly 10% of total DNA content of the genome. Assuming that the members of all gene families comprise about 3/4 of the MR DNA fraction (about 12%), the proportion of the "gene family DNA" is about 9% of total DNA in the genome of *D. melanogaster*. Thus the "minimum" amount of DNA included in genes of all types is about 20% of the total DNA content.

There are no facts allowing us to estimate the "maximum" proportion of gene DNA. The proposals that the interbands of polytene chromosomes actually represent active genes[9-13] may be used as a basis for some estimates. As mentioned before (see Volume I, Chapter 8, Section V and Chapter 12, Section IV) there are approximately 4000 interbands in the polytene chromosomes of *D. melanogaster* which may be long enough to act as a coding sequence. Presuming that the average length of those interchromomeric DNA regions is 1.5 kb, the proportion of DNA included in them is about 4% of the total genomic DNA. Thus a larger estimate may be presented according to which the proportion of "gene" DNA including the "chromomeric genes", "interchromomeric genes", and the "gene families" comprises about 24% of the total DNA and roughly about 30% the "potential gene DNA" (i.e., the DNA classes SC and MR) of the genome of *D. melanogaster*.

II. CHROMOSOMAL LOCALIZATION OF GENES IN *DROSOPHILA MELANOGASTER*

The theoretical basis and principles of genetic mapping of genes have been elucidated e.g. by Haldane,[14] Emerson and Rhoades,[15] Mather,[16] and Judd.[17] The chromosomal mapping of genes in *Drosophila* is mainly based on structural rearrangements detectable by the comparison of synapsed homologs in the salivary gland cells. The accuracy of chromosomal localization certainly depends on the resolution of the optical examination, i.e. capability of recording even the minor details. It also depends on the nature and size of the rearrangements in chromosomes. In the organisms like *D. melanogaster* the occurrence of polytenized interphase chromosomes in easily preparable tissues, like in the larval salivary glands, offers a great advantage to map even the smallest changes in chromosomes. For an optical determination of the chromosomal site of a gene the shortest detectable deletions or duplications are most informative if the gene is known to be included in the area of the rearrangement.[4-7,17] But larger chromosomal changes may also be useful if one of the breakpoints is located close to or even in the area of the localized gene.[1,2]

New molecular methods to insert genetic material in genomes[18] and to link together DNA fragments[19] from different sources have made it possible to amplify DNA sequences isolated from eukaryotic genomes[20,21] e.g. from the genome of *D. melanogaster*.[22] The method for chromosomal localization of labeled DNA by hybridizing it "in situ" into the polytene chromosomes[23] has greatly extended the chromosomal mapping of DNA fragments obtained by using different methods.[22,24,25]

The recognition sites of different restriction enzymes can also be mapped in the chromosomal DNA and used as molecular landmarks. The restriction maps can be compared between the strains and species of *Drosophila*. To create the most detailed maps the DNA fractioned by the restriction enzymes can be completely sequenced.[26]

Restriction enzyme analyses on the overlapping segments of cloned DNA has made it possible to use a new technique called "chromosome walking", i.e. directional analysis and mapping of longer stretches of DNA derived from identifiable regions of polytene chromosomes.[27] By using the previously known chromosomal rearrangements, particularly inversions, for "chromosome jumping" the walking can be continued through the other end of rearrangement into a new region far from the original initiation site of the walk without laborious analysis of the regions between.[28]

By screening the DNA clones against the purified samples of messenger RNAs isolated from the cells of *Drosophila* it has been possible to recognize and ligate DNA fragments carrying specific genes.[29] The sequences of DNA complementary to the messenger molecules can be demonstrated by using R-loop mapping.[30] This technique allows electron microscopic visualization of the size and location of final coding regions, the exons, which appear as heteroduplexes formed by the single-stranded messenger RNA and the complementary sequences in cloned DNA.

By using the methods introduced above, a great number of genes and other sequences of genomic DNA have already been mapped in the salivary gland chromosomes of *Drosophila*. In the following, some examples of recent studies concerning the localization of genes in polytene chromosomes of *D. melanogaster* are going to be briefly reviewed. Table 1 in Chapter 16 includes map distances and/or chromosomal sites of a number of genes localized most accurately in the salivary gland chromosomes of *D. melanogaster*.

A. The X Chromosome

The map compiled on the basis of electron micrographs taken from the thin sections of squashed X chromosomes of *D. melanogaster* differs from the revised reference map of Bridges[31] principally in number of so-called doublets (see Chapter 15, Section II, Revised reference maps). In the EM map of salivary gland X chromosome 139 bands are altogether interpreted to represent only single bands although they have been depicted as doublets by Bridges in his revised light microscopic map.[31]

Thus the total number of separate bands in the EM map is 979 for the salivary gland X chromosome. Since the EM analyses carried out in the future by using more stretched material may be capable of recovering more double bands than it was depicted in the present division maps the total number of bands may still increase. The potential number of bands in the X chromosome, if all the Bridges' doublets could be verified, may be about 1120, which probably is close to maximum number of interphase chromomeres in the X chromosome. Total axial length of the bands of X chromosome indicates that DNA content per chromatid is roughly 22,850 kb, presuming that a band length of about 95 nm corresponds to DNA length of ca 21.6 kb, which is an approximated average DNA content of interchromomere-chromomere units in whole genome of *D. melanogaster* (see Volume I, Chapter 4).

Cytogenetic mapping of radiation-induced lethals[32] has shown that the distribution of vital genes is not equal in all parts of the X chromosome. Certain regions like 1B, 1F-2A, 3AB, 10A, 11A, and 19E-F in the X chromosome are more densely populated with vital gene loci than the other regions in general. On the other hand, some regions like 6E-F and 10B-E seem to contain fewer vital loci than other regions on average.[32]

1. The Zeste-White Region

The subdivisions 3A and B, better known as the *zeste-white* region of X chromosome, have been analyzed most extensively.[6,7,33,34] In the area of the X chromosome starting from the *giant*-locus and ending at the *Notch*-locus altogether 24 loci have been mapped. In the salivary gland X chromosome this area corresponds to the bands between 3A1 (*giant*) and 3C7 (Notch). As it shown by the EM analyses of this region[35,36] and the EM maps,[37] this region is composed of the following sequence of bands (or chromomeres): 3A1-2, new band, new band, 3, 4, new band, 5, 6, new band, 7, 8, 9, 10, 3B new band?, 1, 2, new band, 3, 4, new band, 3C1, 2-3, 4, 5-6, 7 (Figures 1 and 2). Thus the region consists of at least 27 bands, most probably of 28 bands, because the band 3B1 seems to be composed of two subunits, one of them carrying the *period* locus (see Section II.A.7).

Besides the spontaneous mutants known from the region, it was saturated by the mutations induced by chemical mutagens (EMS and NNG) or by irradiation with X-rays.[6,7] More than 120 single-site mutations obtained in the experiments were localized cytologically by using deficiencies known from this region. Crosses were made between the deficiency stocks and the strains carrying the point mutations. The lack of complementation in the hybrids carrying a certain deletion and a point mutation was interpreted as an indication that the mutation in question is located within the area of deleted region in the X chromosome. The linear order of loci was determined from results revealed by the crossing experiments with a long series of partly overlapping deletions. The recombination map showing the cross-over distances between the complementation loci indicates existence of rather regular intervals between the loci, resembling the banding pattern seen in this region in the salivary gland X chromosome of *D. melanogaster*[6,7] (see Figure 2).

In general, the correspondence between the bands and complementation groups is good in this area of the X chromosome, particularly, if it is accepted that the "minibands", one of them following doublet 3A1-2 and the other next to the band 3A6, and the tiny bands 3A5 and 3C4 depicted already by Bridges,[31] are geneless; i.e. the chromomeres of those extremely narrow bands are too small to contain detectable genetic functions. Accordingly, the very heavy doublets of the region, 3C2-3 and 3C5-6, seem to be sparsely populated with genes. The late-replicating chromatin next to the band 3C4 especially seems to be geneless or genetically "nonessential".[38]

Lethals affecting the vitality of flies in different developmental stages have been detected

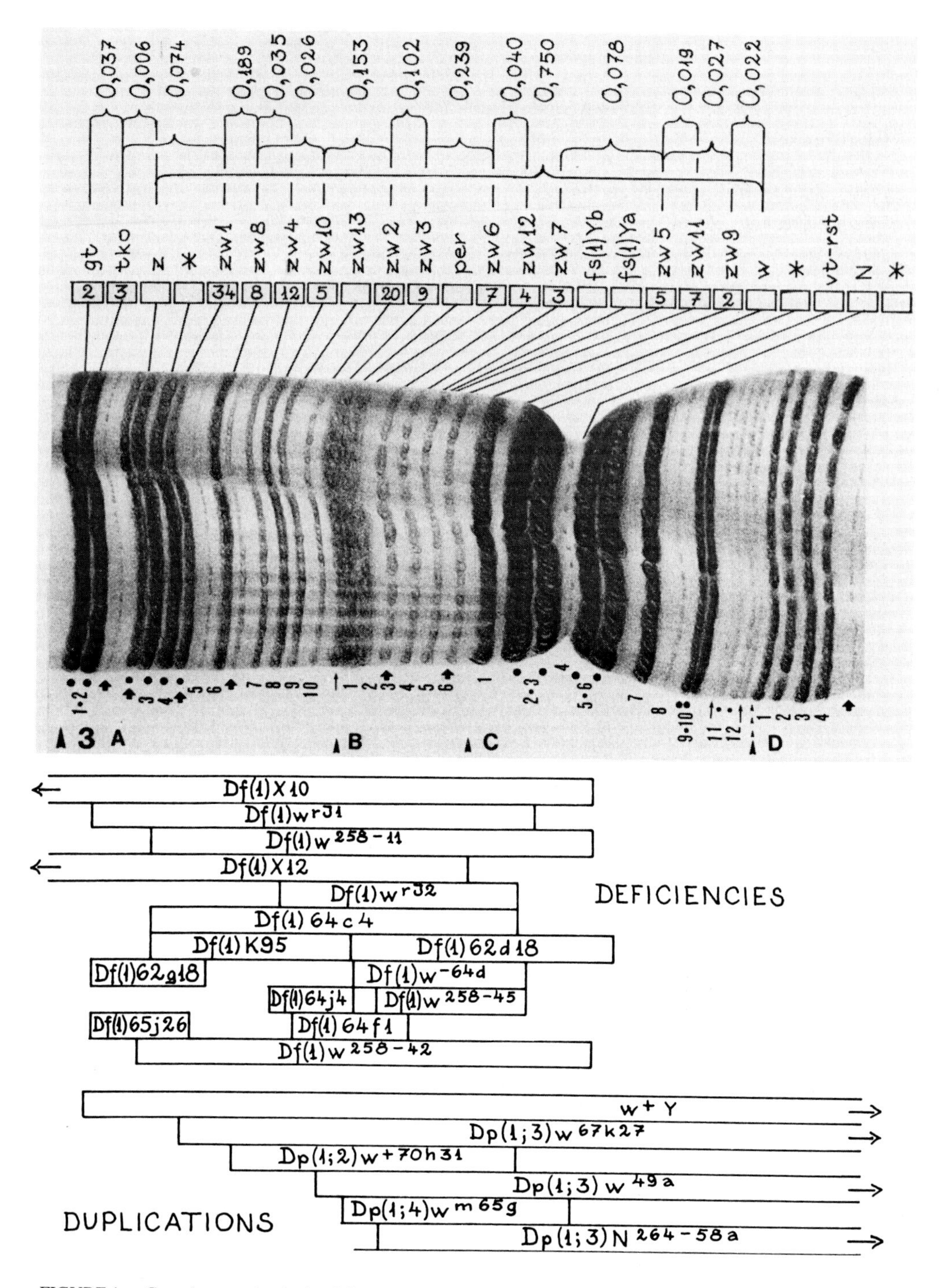

FIGURE 1. Complementation loci and the electron microscopic map of the *zeste-white* region of the X chromosome of *Drosophila melanogaster*. Above the EM-map: the genetic loci identified and designated from the region giant-Notch, with some cross over distances and numbers of mutations localized in the same complementation groups. Below the EM-map: a schematic representation of the approximate sites of some of the deficiencies and duplications used for the localization of induced and spontaneous mutations detected from the region. Redrawn and modified from articles of Judd et al.[6,7]

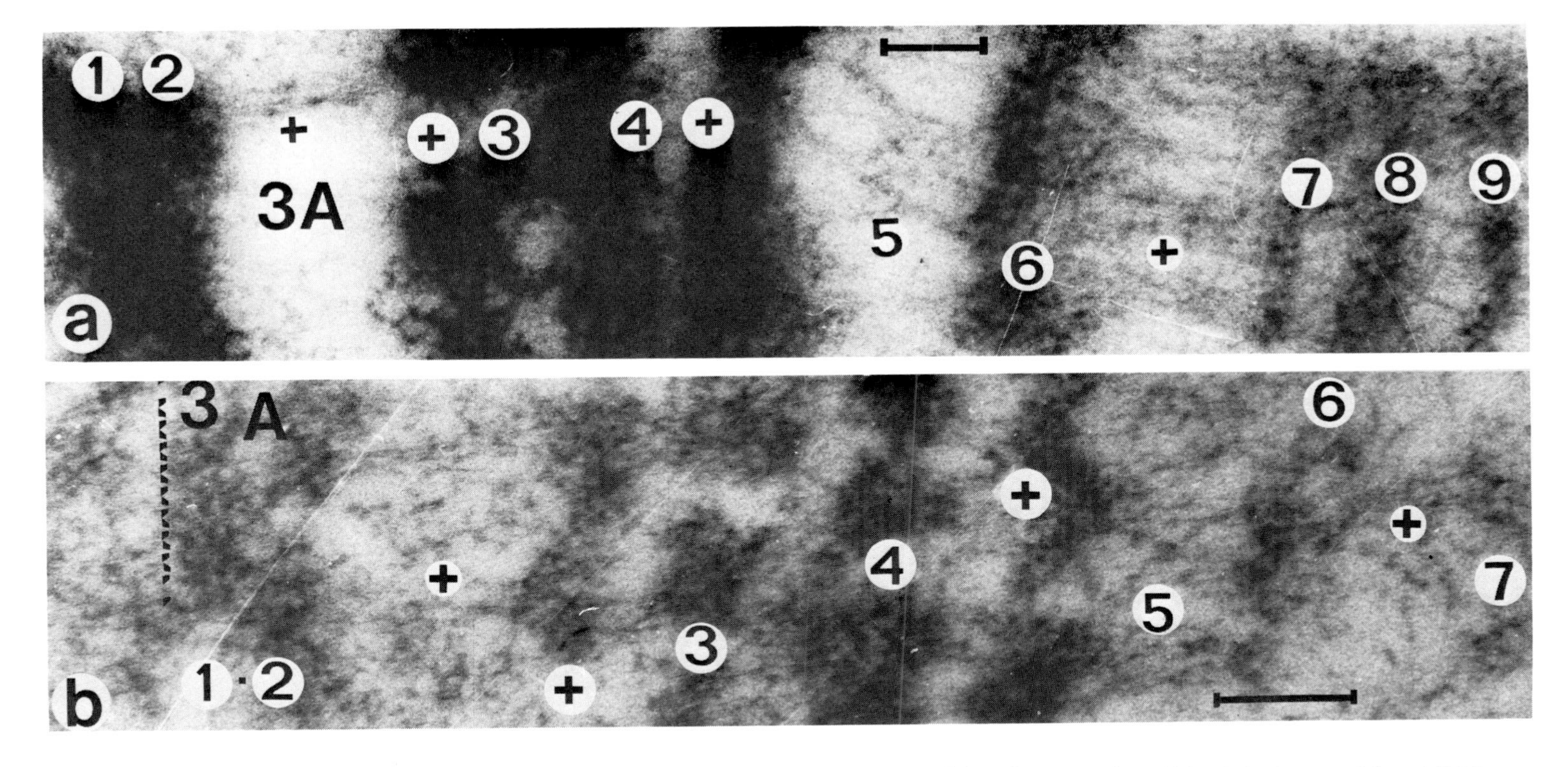

FIGURE 2. Electron micrographs of the subdivisions 3A and 3B of the X chromosome of *Drosophila melanogaster* show 14 bands in the area of the subdivision A and 6 or 7 bands in the subdivision B. In most of the electron micrographs the band 3A4, and in some thin sections also the band 3A3, both of them depicted as singlets by Bridges, are clearly divided into two adjacent subbands appearing as doublets (a and b). The new bands, following the bands 3A2 and 3A6 as well as the Bridges' band 3A5 are extremely faint and sometimes difficult to recognize even in the EM. The band 3A10 is usually well detectable, although occasionally very close to the next band 3B1. The complex of bands 3B1 and 2 may contain a third subunit thus corresponding to the three genetic loci detected in the area (c and d). Appearance and mutual distances of almost equal bands 3B 3-6 may somewhat vary in preparations. Scale bars represent 0.1 μm. (See also Figure 17 in Chapter 8 in Volume I and Figure 1 in this chapter).

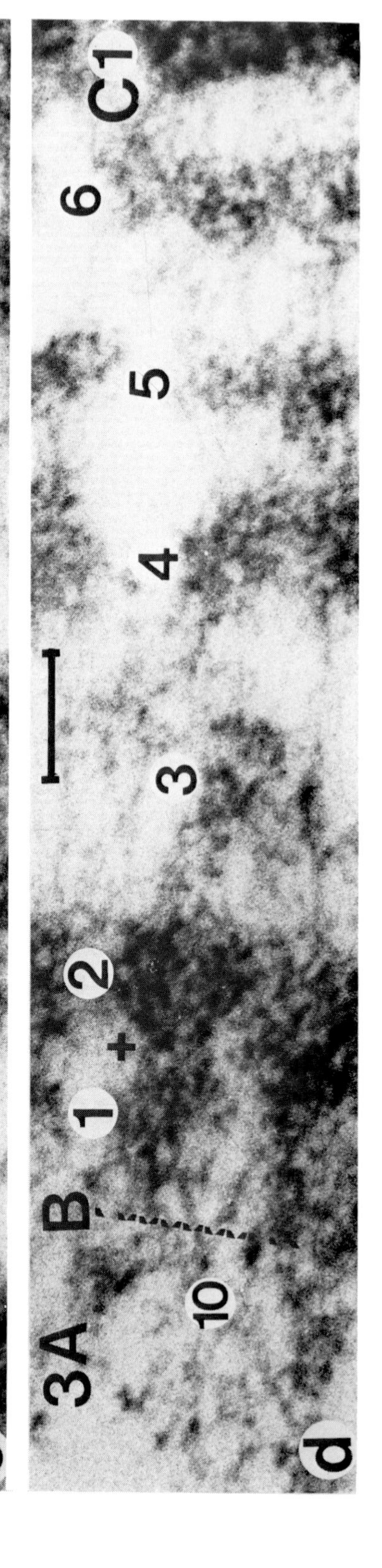

FIGURE 2 (continued)

in most of the complementation loci. Only the loci; *zeste, white, roughest,* and *"verticals"* seem to lack lethal alleles. By complementation experiments it has been possible to determine certain regions e.g. in the subdivisions 3A, 3B, and 3C which apparently contain a considerable amount of "nonessential" DNA.[38] One of those regions is located between the adjacent loci of *zeste* and *zw1*, another is located between *white* and *roughest.* Both of these regions, particularly the latter one, seem be composed of prominent bands. The third region mentioned to contain "nonessential" DNA is located between the complementation loci designated as *zw7* and *zw5* in the area of subdivision 3B. Since this latter region in polytene chromosome X seems to consist of four adjacent, rather similar bands, the closer localization of the nonvital sequences is not evident. However, the chromomeres forming the bands 3B 3-6 contain approximately 15 to 20 kb of DNA per each,[35] which apparently is about 4 to 6 times the DNA content of the gene loci allocated in those bands.

Particular attention has been paid to the genes *zeste* and *white.* Besides the well-described but still puzzling interaction of these loci, certain alleles of *zeste* locus seem also to interact with some alleles of bithorax complex in the chromosome 3R. An interesting feature in the interaction of *zeste* and *white* is the close proximity of two normal *white* loci needed for expression of the *zeste.* In the single X chromosome of males the expression of *zeste* eye-color involves a duplication of the proximal part of *white.* In the females, close pairing of these sites of both homologs is required for the expression of *zeste*-eyes.[39]

Data obtained from extensive experimental studies for elucidating the mechanism of the *zeste-white* interaction indicate that, actually, the *zeste* locus by some way represses the activity of the *white* locus, but the repression is possible or active only if two normal *white* loci are paired or physically adjacent in the genomes.[39] In other words, in females with a *zeste/zeste* combination, the normal function of $white^+$ locus (red eye color) is repressed by the product of *zeste,* if the $white^+$ alleles are closely paired, but not repressed, if the $white^+$ alleles are located separately.

Some of the new revertable mutants of *zeste,* probably arisen as result of insertion of extra DNA in presumptive control region of the locus, seem to code for an active antagonist of the product of the $zeste^+$ allele. These new mutants are capable of expressing the *zeste* eye color even in presence of a single, unpaired $white^+$ gene.[40] From these and other results it has been proposed that the eye color phenotype depends on the balance between the products of *zeste* alleles and the products of $white^R$ (i.e. the proximal region of *white* locus).[40] Interaction of *zeste* and *white* appears although the *white* genes are located in the transposing elements.[41]

2. *The White Locus*

The *white* locus itself has been localized in the left border of the double band 3C2-3.[1,2] The white gene is well known for its high number of mutations and for some highly mutable alleles of it. In 1967, Green[42] proposed the presence of "foreign" elements comparable to the controlling elements of maize[43,44] in certain frequently mutating alleles of *white.* Green was also able to demonstrate transpositions of controlling elements of the *white* gene.[45]

Results obtained from extensive cytogenetic analyses of male-viable and lethal deficiencies[46] have supported the early proposal of Bridges[47] that the heavy doublets 3C2-3 and 3C5-6 may represent a duplication. The finding of a faint band between 3C1 and 3C2 in connection of a deletion in the doublet 3C2-3[46] has been interpreted that actually the *white* locus occupies a separate band close to the left end of doublet 3C2-3.[48] A separate band including the normal $white^+$ function has been demonstrated both by means of light and electron microscopy[2,49,50] at the site of 3C2 in an inversion of the X chromosome. In this special inversion, designated as $In(1)z^{+64b9}$ by Green, the left break point appeared to be located within originally triplicated proximal part ($white^R$) of the *white* locus. In this case the band containing the *white* obviously includes also some chromomeric material from the band

12B9 at the right break point of inversion and the cytogenetic evidence supports the idea that the distal part of band is a fraction of the band 3C2 (Figure 3). Nevertheless, it does not exclude the possibility, that the *white* locus may appear as a separate chromomere in the cells in which the *white* gene is transcribed since the promoter has been mapped in the right end of the gene.[51]

Demonstration of the existence of large transposing elements including the loci *white* and *roughest,*[52] isolation of a recombinant plasmid carrying this huge element,[53] and cloning of it[54] have opened new possibilities to molecular analysis of the *white* region.[55] An insertion of *copia* sequence in $white^{apricot}$ allele and of a foldback (FB) sequence have been detected in the transposing element (TE) *white-roughest.* The latter one, apparently, is responsible of transpositions of the whole element into new chromosome sites. The chromosomal location of the w^a of the TE has been proved by *"in situ"* hybridization of the cloned *copia* sequence to the salivary gland X chromosome.[56,57]

3. *Transposable Elements Within the White Locus*

The physical map of the white locus has been constructed by the restriction enzyme analyses of cloned DNA derived from the area,[58] and later on by direct isolation and microcloning of this region from the salivary gland X chromosome.[51] Aside from the mutant $white^a$ shown to include a *copia* sequence, also some other mutations proposed to include extra DNA[42,45] like $white^{ivory}$ (w^i) and $white^{crimson}$ (w^c) have been subjected to extensive molecular analysis.[59-61]

A tandem duplication of a segment of about 2.9 kb within the *white* locus seems to be responsible for the mutability of the $white^{ivory}$ allele. The highly mutable gene $white^{crimson}$ seems to contain an FB type of mobile DNA element inserted in the duplication segment derived from the w^i mutant. The analyses of the wild-type revertants of w^c have indicated that the wild phenotype is the result of an excision of both the inserted FBw^c element and of one copy of the duplicated sequence (of w^i).[60]

The new dominant mutable allele $white^{DZL}$ (dominant zeste-like) has been localized approximately from the right end of the *white* locus.[62] An insertion element, even larger than that found in w^c, has been demonstrated from $white^{DZL}$. This insertion of about 13 kb of extra DNA within the normal sequence of DNA in the region proximal from the *white* locus seems to cause the high instability of $white^{DZL}$.[63,64] The revertants of w^{DZL} towards the wild phenotype also in this case appear to be resulted in by the excisions in the extra DNA. The excisions of the insertion element, however, do not remove the inserted DNA precisely, but are mostly leaving about 2 to 6 kb long remnants of it. Only the central single copy segment (about 6 kb) is always excised in the revertants. These are able to mutate further indicating that even the remnants of the insertion element are capable of promoting new mutations.[63] Apparently, depending on how much DNA of the inverted terminal segments is still present in the X chromosome of revertants, they have a different stability against new mutations.[42]

In both of the mutable alleles w^c and w^{DZL} the insertion element seems to share homology with the sequences belonging to the dispersive and moderately repetitive FB (fold back sequence) family. In the insertion DNA of w^c there is a single FB element, and a central single copy segment of about 4.0 kb. In the larger insertion element of w^{DZL} there are two terminal FB sequences and the central single copy segment of about 6.5 kb.[64]

The FB element differs from other transposable elements by its long terminal inverted repeats of fold backs. It has been suggested that the insertion of transposable elements carrying the FB components may generally cause easily breakable sites in chromosomes. The hypothesis is supported by the observations that many of the new mutations induced experimentally in the w^c apparently are chromosomal rearrangements starting at w^c.

DNA of the *white* region has been cloned by using several different methods.[54,56] A new technique for microcloning of a chromosome region directly from the squashed salivary

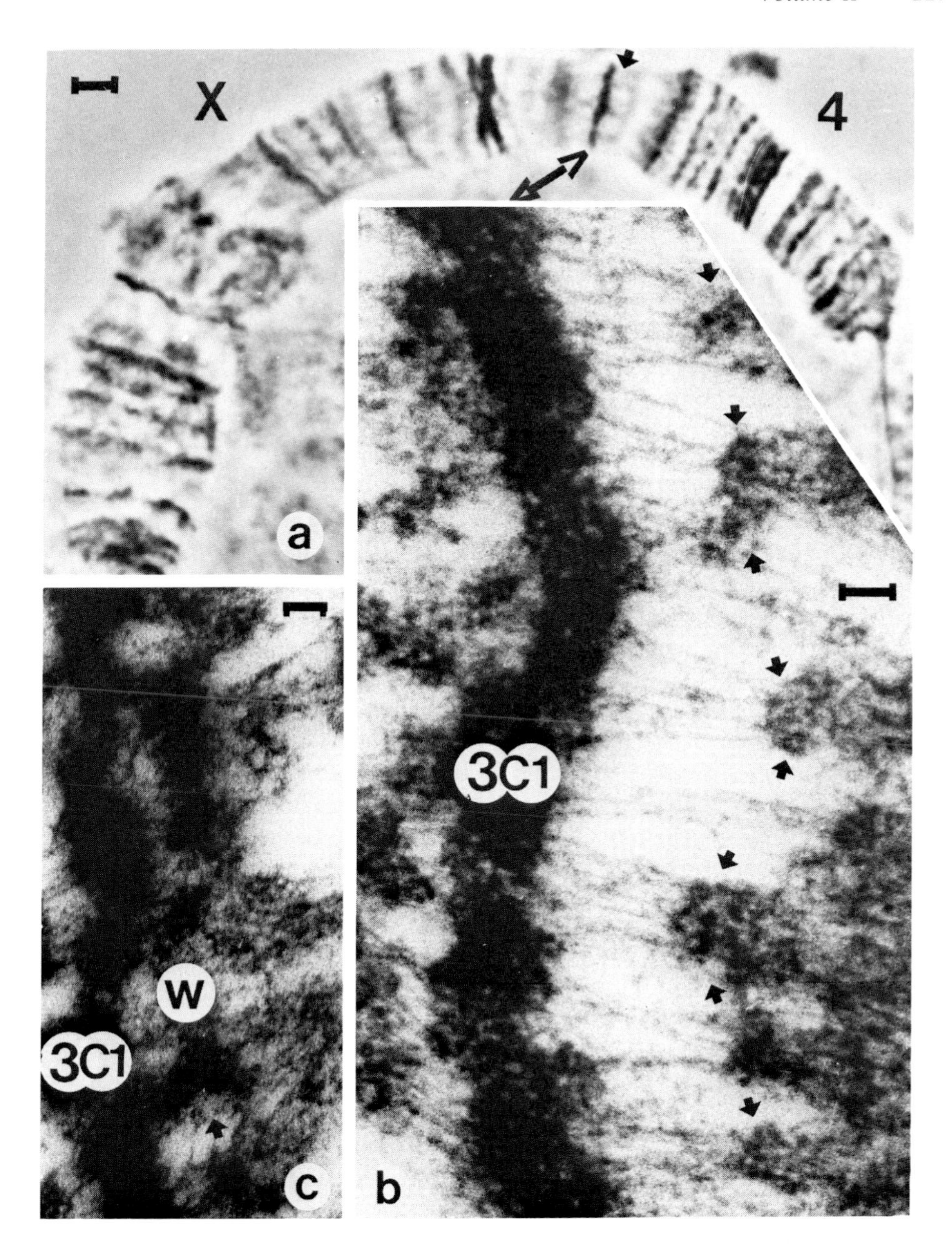

FIGURE 3. An example of using chromosomal rearrangements for the chromosomal localization of genes. A light micrograph illustrating the translocation $y^2 In(1)z^{+64b\ 9}w^{-70\ 1\ 26.5}spl\ T(1;4)$ between the X and the fourth chromosome in *Drosophila melanogaster* (a), and an electron micrograph (b) showing more closely the band 3C1 and fraction of band 3C2 (indicated by arrows), that still includes at least part of the *white*-locus DNA.[2] Electron micrograph (c) and the light micrograph (d) show the band sequence at the area of 3C1-3C2 in the inversion $zeste^{+64\ b\ 9}$ before translocation. DNA of the faint band (pointed by arrows) in the inversion $z^{+\ 64\ b\ 9}$ still includes a normal $white^{+}$ function.[2] By using these and several other rearrangements[35] having their break point in the area of the band 3C2, the *white* gene of *Drosophila melanogaster* has been localized in a distal fragment of the band 3C2. Scale bars represent 1 μm in the light micrographs, and 0.1 μm in the electron micrographs.

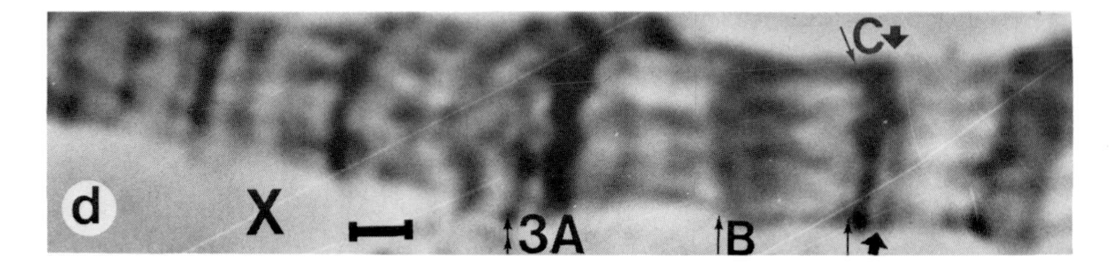

FIGURE 3 (continued)

gland chromosomes by using microdissection has also been applied to study the molecular organization in the *white* locus of *D. melanogaster*.[65] The segments of DNA cloned from the region 3B1-3C2 comprise about 200 kb including about seven to eight bands. Analysis of the microcloned *white*-locus DNA has been carried out by using the mutant w^a known to contain an identifiable copia sequence.[65] Northern blot hybridization analyses of the transcripts of *white* gene isolated both from the wild type and from several mutants of *white*, and from the *zeste* mutant have greatly elucidated the molecular structure of the white region.

The actual transcription segment of *white* locus seems to consist of five exons separated by four introns. The total length of exons is about 2.6 kb. Later studies of white transcripts isolated from pupae and adults have revealed the total length of about 2.7 kb. The transcripts seem to contain a very small proportion of poly(A)RNA.[66] The small exon at the proximal end of the transcript seems to be separated by a long intron (about 2.8 kb) from the other parts. The long intron apparently corresponds to the mutationally silent region found in the *white*-locus. It seems that the DNA of introns can be increased by insertions of transposable elements. Presuming that the length of introns is proportional to the occurrence of crossing overs within the same region, the insertions and excisions of extra DNA in introns seem to regulate the intragenic crossing over.

4. Transpositions of White Locus

Isolation and characterization of another transposable element belonging to the P family of dispersive MR DNA from the *white* locus of *D. melanogaster* indicates that the insertion could be a highly sequence-specific event.[67] And respectively, the excision of P elements seems to be very precise, restoring the original sequence of DNA.[67] A 12-kb-long DNA segment including normal *white*$^+$ function has been used for the P-element-mediated transformation of germ line of *white* mutants in *D. melanogaster*.[68] Experimentally induced transpositions of *white*$^+$ locus into numerous new sites in chromosomes have shown that generally the function of gene stays normal. The P-element-mediated transpositions of many other genes have revealed similar results showing very few recognizable changes in the function of genes.

Besides the higher mutability within the genes, the insertions of transposable elements obviously are capable of increasing the frequency of chromosomal rearrangements in general. In certain cases, deficiencies and duplications have, apparently, taken place between the insertion sites of homologous transposable elements. Thus these elements may have an important role in the induction of local changes in the number and sequence of genes.[69]

In a recent review of spontaneous transitions of the cytologically observable transposing element TE *white-roughest* 41 cases of more than 100 transposition sites have been listed in which the new location of TE has been verified both genetically and cytologically.[70] The genetic[71] and molecular analyses[53-55] of the insertion sites of transposable elements carrying the *white-roughest* region of X chromosome have indicated that the bands at the insertion sites of this element show homology with certain sequence (FB) of elements. When linked to a certain site in the chromosome showing homology with the FB sequence, the TE white-roughest can be cytologically detected due to its large size; e.g. the TE36 has been localized at the proximal part of subdivision 35B in 2L. The excision of TE36 seems to retain the homology with the FB sequence and leave the insertion site without changing the function of genes.[71] These observations strongly support the earlier evidence suggesting that the FB sequence has the principal role in insertion and excision of TE *white-roughest.*

5. DNA Sequence of the White Locus

A remarkable step towards the better understanding of the structure and function of the *white* locus and of its product has been the complete sequencing of the DNA of the area including this genetically and cytologically so well-characterized gene. Besides an exhaustive analysis of the wild-type gene *white*$^+$, O'Hare et al.[3] have also analyzed several rearrangements from the normal sequence (Table 1).

A sequence analysis of the proximal (promoter) end of the gene indicate that the 5′ end of the first exon is located around 3750 base pairs upstream from the starting point of the sequence map which was the first or most proximal nucleotide in the short (5 base pairs) duplication detectable at the insertion site of the *copia* element of *white apricot* mutant. The nearest match with the important TATA-sequence was found at +3772 and thus the transcription of the gene was proposed to initiate at about +3737 and the translation of messenger at about +3511. The base sequence (AATAAA) usually preceeding the polyadenylated end of the messenger was found at the position −2207 indicating that the 3′ end of *white*$^+$ RNA is located around the site −2230. Thus the total length of w^+ RNA is about 5.97 kb, and the length of w^+ messenger after splicing the four introns is about 2.6 kb.

The insertion sites localized within the intron areas of the gene seem to induce mutations showing reduced amounts of transcripts rather than having effects on the coding sequence or on the splicing of the final messenger. The size and the base sequence of w^+ messenger-RNA indicate that the properties of the translation product of *white* locus have resemblance with those of membrane proteins.[3] As it was pointed out by the authors[3] the knowledge of the DNA sequence of the *white* locus may greatly advance the solving of many problems concerning the structure and function of the *white* gene and of its product, the mutations of locus as well as the interaction of *white* and *zeste.*

6. The Notch Region

The deletion mapping of X chromosome has localized two genes: *roughest (rst)* and *verticals (vt)* between the loci of *white* and *Notch.*[72] The *rst* locus apparently is very close to the distal border in 3C5 or in the tiny band 3C4. The latter possibility seems, however, less probable since the DNA content of the chromomeres of band 3C4 must be very low. The *vt* locus appears to be within the heavy doublet 3C5-6.[72]

The sex-linked dominant gene, *Notch,* which is lethal as homozygous and as hemizygous but appears in the heterozygous females *N*/+ as typically notched wings, has been localized in the band 3C7. The chromomeres of this band may contain roughly about 45 to 50 kb of DNA. Besides several alleles of *Notch,* the same band seems to house alleles of another dominant gene, *Abruptex (Ax),* and alleles of *facet (fa), split (spl),* and *notchoid (nd).*[73] A long series of extensive cytogenetic analyses of the *Notch* region have been published by Welshons[73] with his collaborators[76-80] during the last 20 years. From results of those studies it seems evident that the *Notch* locus occupies the whole length of the band 3C7. Several *Ax* alleles have been mapped within the area of *Notch* locus[81] (Figure 4).

Apart from the well-known morphological effect of the *Notch* gene on the development of wings, recently some biochemical defects have also been recorded in connection of certain deficiencies in the *Notch* region. Results revealed by the studies on biochemical processes proposed to be encoded by the *Notch* alleles indicate that *Notch* has a wide spectrum of effects on the respiratory chain of mitochondria. The six visible mutants (fa, fa^g, fa^{no}, spl, nd, and nd^2) seem also to have an effect on activity of at least four different enzymes.[82]

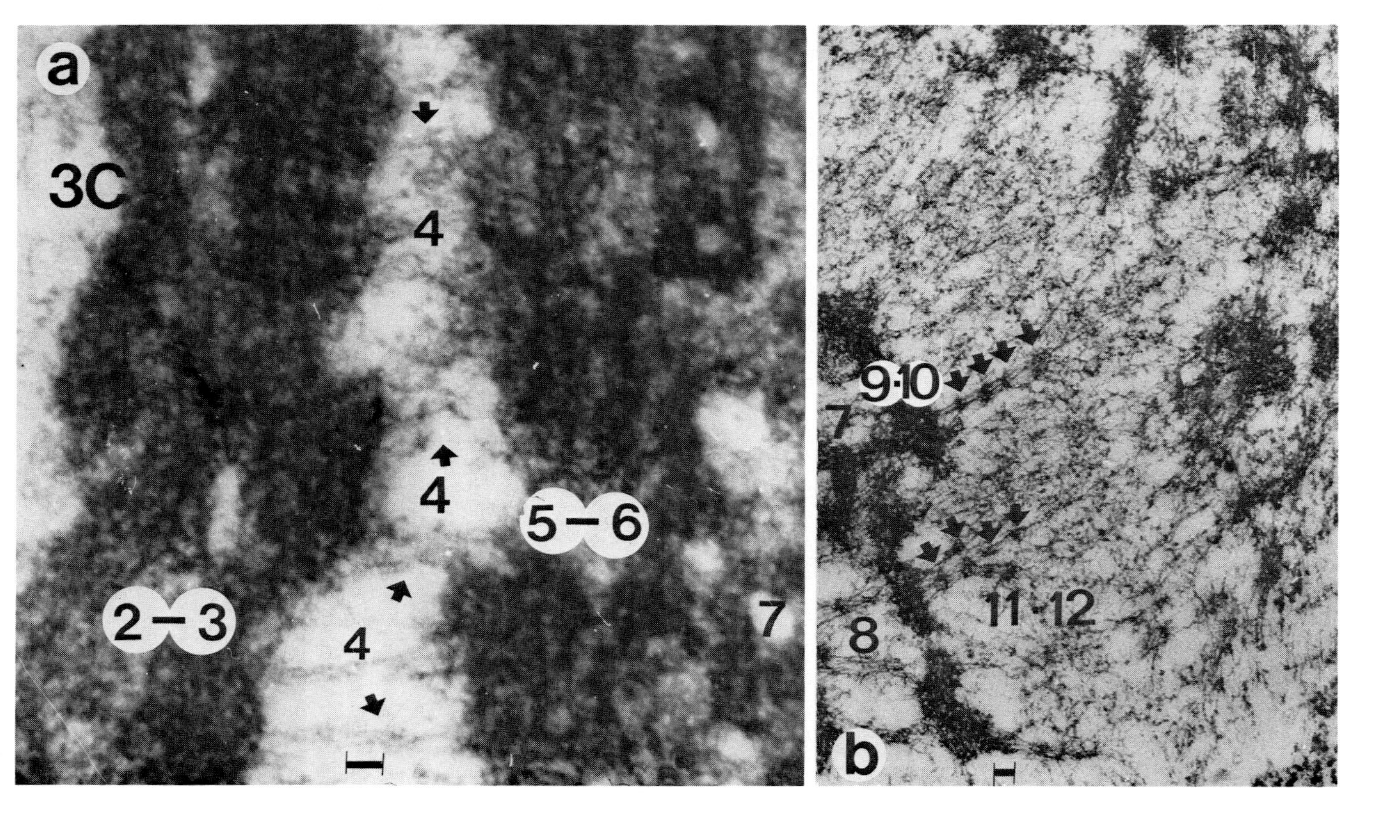

FIGURE 4. Electron micrographs of the X chromosome bands including the *white* and *Notch* loci as well as the glue protein gene *sgs-4*, and eye color gene *vermilion* in *Drosophila melanogaster*. (a) Electron micrograph of a thin section of heavy bands 3C2-3 and 3C5-6. The faint band 3C4 is usually difficult to demonstrate even in the EM. Approximate site of *white* locus in the band 3C2 is indicated with arrows. (b) Puffed tiny bands 3C11-12 contain the glue protein gene *sgs-4*. Appearance of nonpuffed chromomere groups in thin sections point to certain asynchrony in the activation of region. (c) A thicker section of the same region of X chromosome carrying a genetic triplication in the proximal part of white locus. The triplication appears in the chromosome as an unusual thickness of the doublet 3C2-3. The inversion z^{+64b9}, introduced in the previous Figure 3, was obtained by M. M. Green by irradiating this triplication stock. (d) Electron micrograph showing the band 3C7 including the *Notch* and several other genes. (e) An electron micrograph of the band complex 9F12-10A1-2. The band doublet 10A1-2 also includes several genes,[5] e.g. *vermilion*. Scale bars represent 0.1 μm.

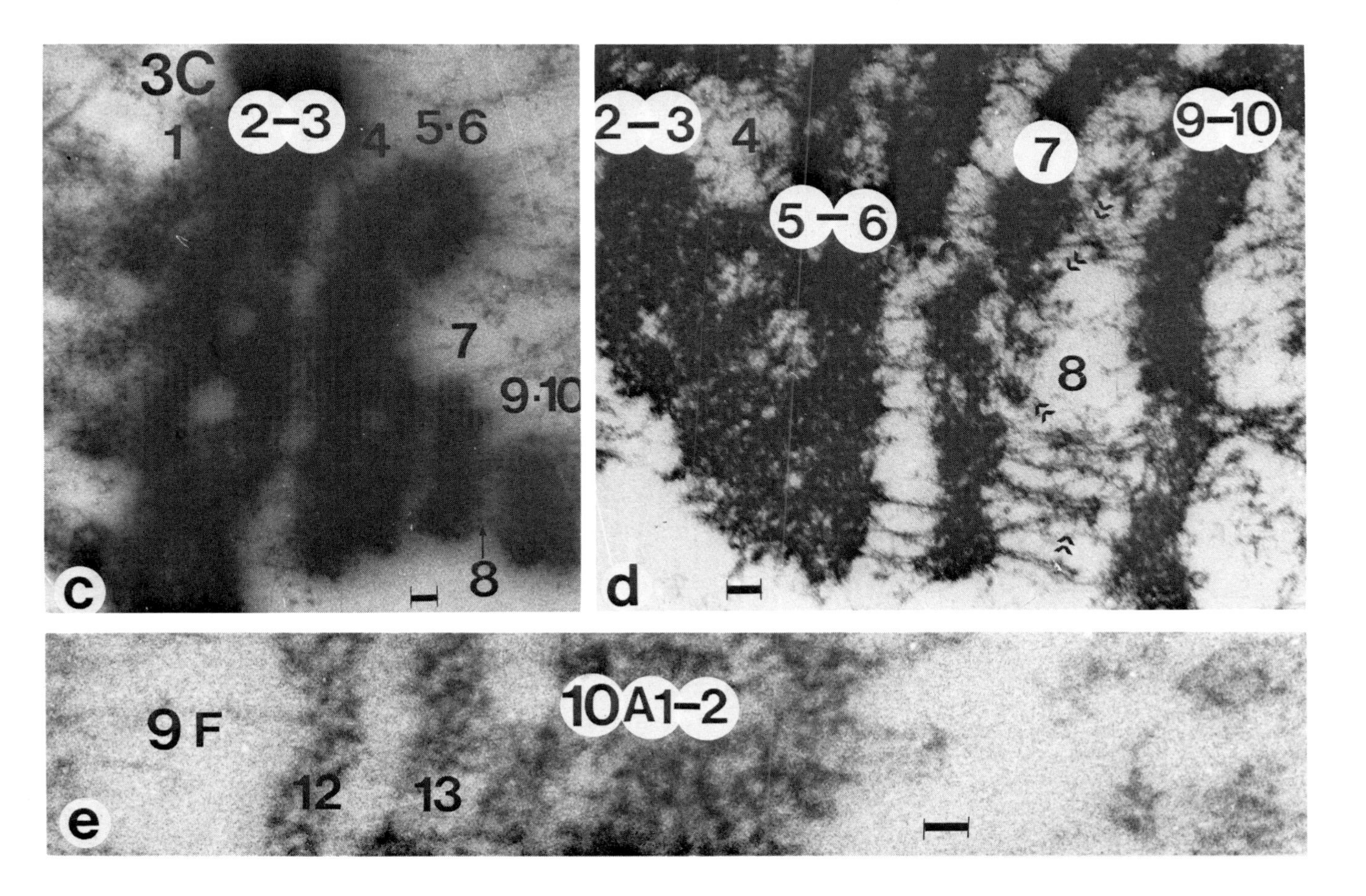

FIGURE 4 (continued)

Results from restriction enzyme mapping of about an 80-kb-long segment of DNA isolated from the area of the band 3C7 of the X chromosome, and from comparison of this segment with a proposed *Notch* poly(A)$^+$ RNA of size class of about 10.5 kb, all indicate that the size of the *Notch* locus is approximately 40 kb.[83] Accordingly, another molecular analysis of the *Notch* region has revealed about a 37-kb-long transcription unit in DNA sharing homology with about an 11.7-kb-long poly(A)$^+$ RNA. Closer analysis of the homological sequences has demonstrated nine coding regions ranging from 0.13 to 7.25 kb in size.[84] Variation in the composition of transcript was found to exist mostly at the 3′ end. Seven insertions were detected by the analyses of 24 point mutations of *Notch* locus. Most of the point mutations were changes within the coding sequences.[84]

By using the overlapping cDNA clones sharing homology with the major embryonic transcript of the *Notch* locus the whole sequence of this more than 10-kb-long RNA has been determined.[85] An 8109-bp-long open reading frame has been found in this long transcript allowing also the prediction of amino acid sequence in potential protein coded by translation. The sequence was compared with different types of proteins. A region of polypeptide, which seems to be composed of 36 tandem repeats of a sequence of 40 amino acids shows resemblance with protein called epidermal growth factor (EGF). From these results it has been suggested that the large *Notch* transcript produced during the embryogenesis may code for a membrane protein having properties needed for interactions between cells.[85]

In the *Notch* locus, as in the case of the *white* locus, the instability of a number of mutable alleles is apparently caused by the presence of the FB type of transposable elements. However, the presence of insertion sequences do not always appear as a mutant phenotype.[86] The transcription product of *Notch* locus seems to contain a specific repeat unit found also in many other developmental loci. This novel family of repetitive DNA is designated as *opa* family.[87] The repeat units belonging to the *opa* family are transcribed and translated and they are principally composed of triplets coding for a specific amino acid, glutamine. Both the enhancement and suppression of a deletion mutation fa^{swb} (strawberry) have found to be in connection to certain chromosomal rearrangements in area of adjoining distal bands of the *Notch*-region.[88]

7. Cytogenetic and Molecular Studies on Other Genes and Regions in the X Chromosome

A series of extensive cytogenetic analyses have been carried out on the region 2B3-2B11 of the X chromosome of *D. melanogaster*, cf. e.g.[89,90] According to the results obtained from these studies the region consists of at least 11 bands and includes the puffed region 2B5-6. The more than 40 mutations affecting on viability seem to form 6 complementation groups. Four of them are apparently locating in the adjacent bands 2B3-4 and 2B5. The two other groups are more proximal.

The *period (per)* locus belongs to the group of genes affecting on biological rhythms in *D. melanogaster*.[91-93] Starting with report of the clock mutants[91] a large number of later investigations have been published around this topic. The *period* gene has been localized in the distal bands of the subdivision 3B.[92] One of the several transcripts of this region is obviously involved in the genetic control of circadian rhythms in *Drosophila*. By using the P-element-mediated transformation of the *period*$^+$ locus into the germline of certain mutants which have lost the normal rhythms it has been possible to return normal functions of period locus, e.g. the short-term oscillation of the courtship song of males.[93]

More recently, the X chromosome region 4BC has been analyzed cytogenetically by using 26 visible rearrangements and four smaller changes in this region.[94] Eight genetic loci (*mei-9, norpA, lac, omb, bi, Qd, rb*, and *amb*) have been localized in the area of bands 4B4-4C8. The narrow Bridges' doublet 4C5-6, seen as two unequal separate bands in the EM-graphs, and the adjoining regions including a new band in EM, seems to contain 4 of the loci (*lac, omb, bi*, and *Qd*).

Characteristic for the mutants of the X-chromosomal locus *cut* are different types of deformations (cuts) in the edges of wings. Some of the cut mutants seem to affect also appearance of kinked femurs, and a whole group of spontaneous and induced mutations are lethal. Chromosome walking analysis of a DNA segment of about 200 kb from the region of the heavy doublet 7B1-2 seems to include the *cut* locus. This locus has appeared to be one of the most mutable genes in the X chromosome of *D. melanogaster*. The molecular analysis of 43 mutants of *cut* locus, belonging to 6 mutant groups according to their effects, has uncovered four segments in this locus which seem to have phenotypic effects in different tissues.[95]

Cytogenetic analyses of the complex band 10A1-2, and the adjacent regions have shown that certain bands may include several gene loci with unrelated functions.[4,5,96] In the heavy "doublet" 10A1-2 at least three gene loci, *vermilion (v)*, the lethal *l(1)BP4*, and *sevenless (sev)* have been mapped and these are intervened by "silent" DNA regions, which actually comprise about 70% of chromomeres in 10A1-2. Three other unrelated functions have been localized in the region around the band 9F12.[96,97] The heavy "doublet" 10A1-2 has been supposed to contain about 200 kb DNA per chromomere, but the chromomeres of much narrower band 9F12 may contain only about 30 kb (see Figure 4).

The resessive gene, *rudimentary*, was one of the first genes mapped in the linkage group of X. However, both the exact chromosomal site and the biochemical background of its functions have long escaped from their students. Cytogenetic localization has suggested the band 15A1 for the chromosomal site of *rudimentary*. According to EM mapping of X chromosome bands the chromomeres of the Bridges' doublet 15A1-2, which is interpreted as single band in the EM map, may contain about 50 kb of DNA. (See EM map, Volume II, Chapter 16, Figure 1).

Molecular analysis of *rudimentary* transcripts has shown that the gene spans a segment of more than 14 kb. The transcript is composed of seven exons intervened by six introns, longest of which in about the middle of locus, obviously divides it into two main domains. These domains, which are split further by shorter introns, seem to code for two enzymes CPSase and ATCase.[98]

The most proximal divisions 19-20 of the X chromosome have also been subjected to several intensive cytogenetical studies, particularly, by using the induced chromosomal deficiencies.[99-103]

B. The 2L Chromosome

Most of the doublets depicted by P. N. Bridges in his revised version of the salivary gland chromosome map of 2L of *D. melanogaster* are also present in the electron micrographs as two separate bands, or by showing recognizable doubleness in their structure. However, 58 of the Bridges' doublets have been interpreted as single bands in the EM material used for drawing the composite EM map of 2L presented in this book. Thus the total number of bands in the EM map of 2L is 869. The potential total (maximum?) number of bands in the left limb of the second chromosome is, if all the doublets depicted in the revised map could be verified, about 927.

As shown by Table 1 in Chapter 16, most of the accurately localized genes in the 2L are mainly located in the distal divisions 21-25, but also in the proximal divisions 34-40. A large scale cytogenetic analysis of 2L has been recently carried out from divisions 23-26.[104] As an example of single loci from the 2L subjected to intensive analyses one may mention e.g. those coding for the enzymes *Adh* and *Ddc*.

1. Gene Loci of 2L Chromosome

In *D. melanogaster* the *Adh* (alcohol dehydrogenase) locus has been localized within the bands 35B1-B3.[105] By using the P-element-mediated transfer of an 11.8-kb-long DNA segment of DNA including a single copy of *Adh*$^+$ locus into the embryos of an *Adh* null mutant it has been possible to increase the ADH enzyme activity in the tissues of transformants. Aside from the successful transfer of cloned active locus of this gene in a deficiency strain of *D. melanogaster*,[106] the DNA of the proximal 35A and distal 35B has been subjected to molecular analysis. The cloned sequence of about 165 kb was reported to include at least three genes: *Adh*, *osp* (outspread), and *noc* (no-ocelli).[107] The alleles of *osp* seem to occupy about a 52-kb-long region of DNA, and apparently on both sides of the *Adh* locus, partly overlapping it. The size of *noc* locus also seems to be very long comprising about 50 kb of DNA.[107] The chromosomal location of *Adh* may be more distal than expected earlier[105] (see Chapter 16, Table 1).[105]

The transformation of genes with P-elements introduced by Rubin and Spradling[108,109] has also been applied to the *Ddc* locus of *D. melanogaster*. The *Ddc*$^+$ gene has been successfully transferred into the embryos of the mutant strain.[110] The cytogenetic analyses on the region including the *Ddc* (dopa decarboxylase) locus have shown that the altogether over 200 mutations fall in 13 complementation loci, with one visible effect (*hook*) and 12 lethals.[111] Most of the genes were localized in the region 37B10-C4 composed of about seven bands. Only one of the loci seems to be included in the proximal bands of subdivision 37C.[111] Thus the distribution of detected genes in the chromosome region including the *Ddc* locus is not even, but on the other hand, the number of loci is higher than the number of bands.

The gene coding for myosin heavy-chain (*Mhc*) in *D. melanogaster* has been recently cloned and analyzed. The results suggest that *D. melanogaster* has only a single copy of *Mhc* gene. The transcribed segment is about 19-kb long, and it includes at least nine introns. Three different mature messenger RNAs have been detected and isolated. These messengers represent molecular lengths of 7.2, 8.0, and 8.6 kb, and they are differently spliced in their 3′ ends. The messenger types are produced or active in different phases of development.[112]

2. The Histone Gene Family

The organization and evolutionary background of the repetitive histone coding genes in *D. melanogaster* have been elucidated in several review articles.[113,114] Chromosomal localization of *histone genes* was first succeeded by using *in situ* hybridization of histone messenger from three species of sea urchins in the salivary gland chromosomes of *D. melanogaster*.[115] The sea urchin histone messenger seemed to hybridize with the DNA of the proximal region of 2L chromosome. The strongest labeling was found over the band sequence 39D3-39E1-2 near the chromocenter.[115] Ectopic pairing of the chromatin of histone locus bands with some other regions in 2L and 3R did not indicate any sharing of histone DNA with those other regions. Apparently, the similarity leading to recognition and ectopic pairing is maintained by other repetitive sequences in DNA of those regions. Thus the *histone-gene family* in the chromosomes of *D. melanogaster* seems to be located as one complex at 39DE of 2L[115] (cf. Figure 5).

Closer comparisons of the reading directions of individual members of histone gene family has shown essential differences between sea urchins and in *Drosophila*.[113] While the sequence (H4, H2b, H3, H2a, H1) of the five histone genes in genome of sea urchin is transcribed into the same direction, in the genome of *Drosophila* both the sequence of family members and the directions of their transcription are different.

The salivary gland chromosome region including the histone genes has been further characterized by genetic analyses of spontaneous and induced changes in area.[116-118]

C. The 2R Chromosome

In the EM map of salivary gland 2R chromosome of *D. melanogaster*, 1009 separate bands have been depicted. If the doubleness of those 143 bands, drawn as doublets by Bridges and Bridges,[119] but interpreted now as single bands in the EM map can later be

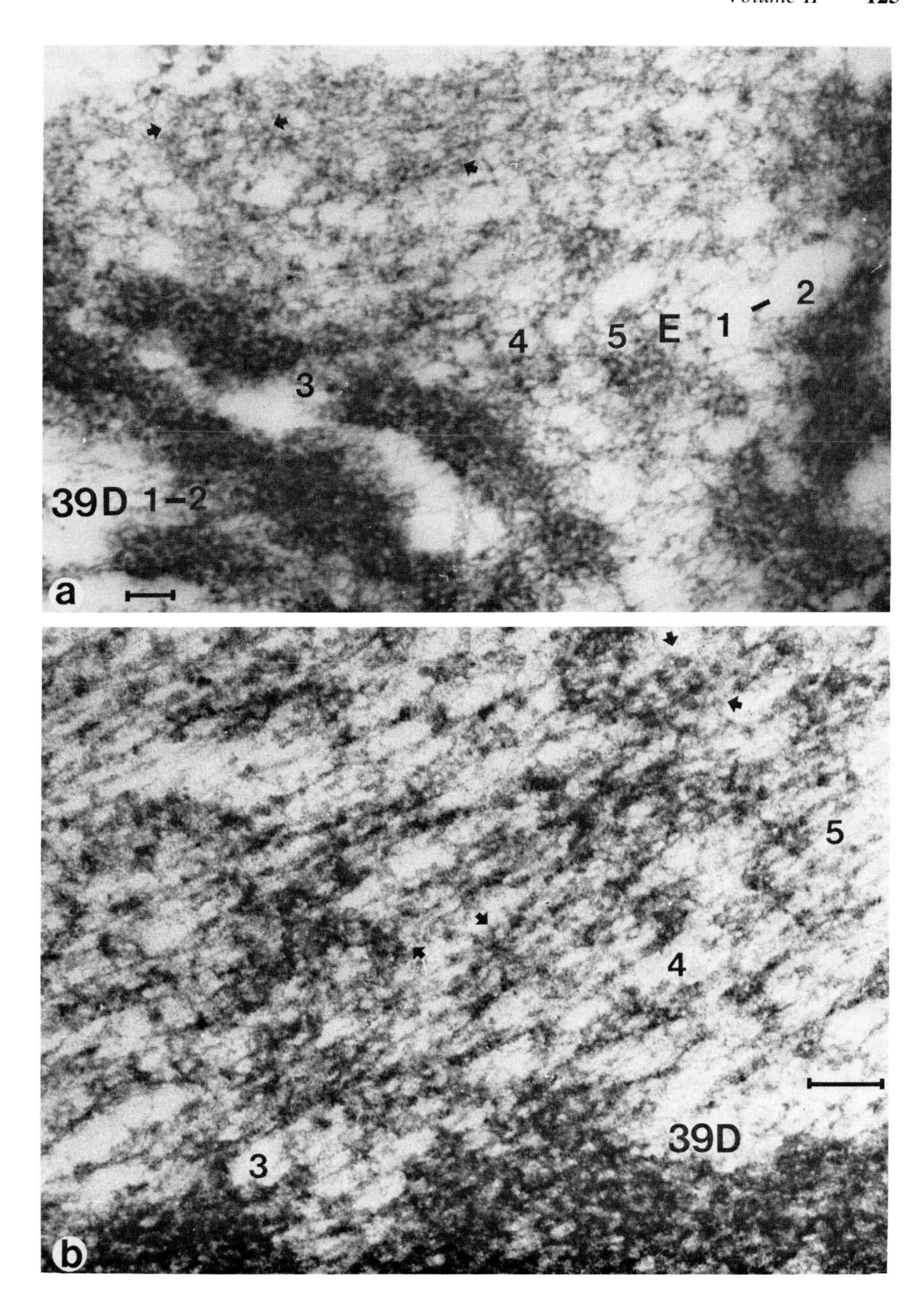

FIGURE 5. Electron micrographs of the region including the histone genes in the salivary gland 2L chromosome of *Drosophila melanogaster*. (a) EM graph of partially activated histone gene locus in the 2L chromosome, showing the main activity in region of bands 39D4 and D5 as well as in the 39 E1. Presumed sites of transcription are pointed with arrows. (b) A higher magnification of a more stretched region of histone genes. (Scale bar represents 0.1 μm.)

proved, the potential (maximum ?) number of bands in 2R would be around 1152. In the right arm of the second chromosome most of the accurately localized genes are in two proximal and five distal divisions (see Chapter 16, Table 1).

1. The 5S RNA Locus

A most extensively studied region in the salivary gland 2R chromosome is the *5S RNA locus* at the subdivision 56F. According to the pioneer studies in the field, the 5S RNA genes of *D. melanogaster* consist of about 200 copies per chromatid, probably grouped in the adjacent chromomeres located as shown by the banding pattern in 56F. The 5S RNA genes were localized in this region by using the RNA-DNA hybridization with autoradiography.[120,121] Localization of 5S RNA hybridized to the induced lampbrush stage of polytene chromosome 2R shows labeling in the chromatin loops protruding out from the chromomeres of the heavy band complex in 56F.[122]

The molecular evidence has later suggested an existence of a single continuous cluster of about 160 tandem repeats in the *5S RNA locus* without any detectable large interdispersive segments.[123] The results from molecular studies on the locus indicate that *5S RNA* genes in *D. melanogaster* form a single continuous cluster.[124] Structural correspondence of continuous cluster and to the presumptive chromosomal site of 5S RNA genes is problematic (cf. EM map, Chapter 16, Figure 1). The total length of 160 units of about 0.4 kb (see below) is roughly 64 kb. In the salivary gland 2R chromosome the complex of heavy bands 56F1-9 contains approximately 360 kb of DNA per chromatid. Hybridization experiments show labeling on several bands containing about 180 kb DNA per chromatid. If the 5S RNA locus is composed of a single cluster the genes should be located in an area of one doublet, probably the most proximal doublet next to the puffed region of 56E which seems to show a most intensive label.

Localizations of the 5S RNA genes have also been carried out in the genomes of some other species of *Drosophila*. The 5S RNA genes of *D. hydei* have been localized in the bands 23B1-2 and adjacent regions in the second chromosome.[125] In the species of the *virilis group* of subgenus *Drosophila*, a variation in the location of 5S RNA genes has been found.[126] From results of other studies it has been suggested that two ancestral loci for 5S RNA genes may exist. In those two lines derived from those ancestral forms the opposite loci have been conserved.[127] Accordingly, in *D. funebris* three distinctive sites of labeling can be observed after RNA-DNA hybridization in the salivary gland chromosomes.[128]

Molecular studies on cloned 5S DNA of *D. melanogaster* have elucidated the sequence organization in the repeat units. A segment of DNA inserted to a recombinant plasmid, used to propagate the 5S RNA genes in *E. coli* cells, contained 32 contiguous repeats of a sequence of about 380 bp. Each of those repeats were shown to be composed by a gene unit of about 120 bp and by a spacer unit of about 260 bp. The cRNA of cloned genes hybridized in the heavy band complex at subdivision 56F, just as the earlier experiments with isolated 5S RNAs had revealed.[129] For wider discussion concerning the structural and functional organization in redundant gene loci the reader is referred to the review in Reference 130.

2. Other Genes of 2R Chromosome

From recent studies on other gene loci of 2R chromosome the molecular analyses of *cuticle* genes may be taken as an example. Cloning and characterization of about a 36-kb-long DNA segment from the region 44D of 2R seems to contain these genes.[131] In a group of five genes expressed at about the same period of larval development of *D. melanogaster* four genes seem to form a cluster within a segment of 7.9 kb. The poly(A)RNAs produced by at least three of those genes, probably by all of them, are coding for cuticle proteins. The fifth gene, located separately at distance of about 8 kb from the clustered genes, appears to be active in late third instar stage.

Sequence analysis of the coding regions has shown that the four clustered genes are identical, including a code for signal sequence, but two of them are transcribed into the opposite direction from the two others. The fifth gene locating separately may be a pseudogene. Thus the *cuticle* genes of *D. melanogaster* seem to form an example of a small gene family with its own phylogeny. Evolution of this gene family seems to contain two gene duplications, an inversion and an unequal crossing over.[132]

D. The 3L Chromosome

According to the interpretation of the electron micrographs of thin-sectioned salivary gland chromosomes the left limb of the third chromosome contains 1032 bands. Since 41 of the Bridges' doublets in 3L have not been verified on the basis of the EM material used for the mapping, the potential (maximum?) number of bands may be 1073 for this arm of the salivary gland chromosomes of *D. melanogaster*.

1. Genes of Chromosome 3L

From the molecular and other recent approaches to localize genetic functions in the 3L chromosome, some examples are presented in the following. DNA from regions of intermoult (glue) puffs in the salivary gland chromosomes of two species of *Drosophila* has been isolated and characterized.[133] In *D. melanogaster*, the isolated DNA is hybridized in the region 63CF. In *D. hydei* the homologous region appears to be in subdivision 90B. DNA derived from the region of an ecdysteroid regulated puff, 74EF, has been cloned.[134] A polyadenylated transcript of about 2.7 kb seems to be complementary with the DNA of this prominent puff. Closer analysis of this DNA segment has made it possible to recognize and localize the typical initiation sequences of transcription including the "TATA" and "CAAT" boxes.

Some of the genes characterized from 3L chromosomes belong to the group having multiple loci in the genome of *Drosophila*. A locus of one of those genes, the salivary gland secretion protein gene *Sgs-3*, has been detected in a segment of the 3L chromosome analyzed recently using numerous small deficiencies induced by irradiation.[135] Several other loci were found from the same segment around the Bridges' division 68 in 3L. The locations of recessive visible mutations in the area are: *rose (rs)* at distal 68A, *rotated* abdomen (*rt*) at 68C, the eye color locus *vin* at 68CD, *approximated (app)* at 69A, and the locus of *eyegone (eyg)* at 69C with the gene *gs*. Aside from these genes 5 lethals were also localized in the region 68A-E. The structural gene coding for a larval serum protein (*Lsp-2*) was localized in the area of bands 68E3-4 with two of the lethals. A structural gene for esterase (*Est-6*) was mapped close to that of the visible mutation *app* at the subdivision 69A.[135]

A segment of DNA cloned from 3L includes the *hairy* alleles affecting on segmentation in *Drosophila* embryos. By utilizing a mutant allele of the *hairy* locus carrying the recognizable gyp*s*y element, a molecular analysis has been done in the segment localized in the subdivision 66D of 3L. The mutant alleles having a *gypsy* insertion are, in general, suppressible by the *suppressor* of the dominant gene *Hairy wing*.[136]

E. The 3R Chromosome

In the salivary gland 3R chromosome of *D. melanogaster*, the electron microscopic mapping has revealed 1147 bands of which 86 have been interpreted as single bands, although they have been depicted as doublets in the revised light microscopic map drawn by P. N. Bridges.[137] Thus the maximum number of bands may be about 1233 in 3R, if these 86 bands are indeed doublets.

Aside from the two large complexes of developmental genes *Antennapedia* and *Bithorax*, the right arm of the third chromosome contains many other gene loci subjected to intensive cytogenetic and molecular studies. For instance, the gene designated as *ninaE*, and coding for the photoreceptor pigment protein, opsin, has been recently localized with a small deletion to the bands 92B6-7.[138] Another opsin gene, designated as λ *Dm rh2*, expressed in other receptor cells than *ninaE* gene, has been localized to the chromosomal position 91D1-2 in 3R.[139] The *rh2* gene encodes a polypeptide that shares about 67% homology with opsin encoded by the *ninaE* gene.

A cloned DNA sequence of about 18 kb, including several genes, some of them active during the embryogenesis, has been localized into the region 99D in distal 3R.[140] The gene *Acph-1*, coding for the enzyme acid phosphatase-1, has also been mapped close to the distal end of 3R chromosome in *D. melanogaster*. Its genetic position is 101.1 and the chromosomal site is at the subdivisions 99DE.[141] Most extensively analyzed has been the region of 3R including the genes coding for the enzymes acetylcholine esterase and xanthine dehydrogenase.

1. The Rosy-Ace Region

Results of the earlier cytogenetic studies on the organization of *rosy* locus at region 87D have been reviewed by Chovnick et al.[142,143] Cytogenetic mapping of *rosy* locus and of adjacent regions in the 3R chromosome by using a great number of induced mutations has elucidated the genetic organization in the region.[144-146] The interval extending from the bands 87D2-4 to the bands 87E12-F1 is composed of about 23 to 24 bands in the revised reference map of 3R,[137] and of about 28 bands in the EM maps. The same interval has been shown to contain about 21 complementation loci.[146]

Closer analysis of the segment of the *rosy* locus proposed to contain a gene coding for the enzyme xanthine dehydrogenase (XDH) has uncovered sites of control elements in the region adjacent to the XDH structural gene.[147] The P-element-mediated transfer of DNA including the XDH gene has shown that the activity of the XDH locus can be close to normal in numerous chromosomal sites.[148]

Molecular mapping of a DNA segment of about 315 kb from the *rosy* region of the 3R chromosome[27,28] includes, apparently, the chromomeres (bands) from the doublet 87D5-6 to the band 87E 5 or 6. According to the revised reference map of Bridges,[137] this segment of the salivary gland 3R chromosome consists of about 13 bands. The EM map shows 14 to 15 bands in the same interval. The approximate DNA content per chromomere varies from about 160 kb to about 7 kb, the lowest DNA content may be about 3 kb.[27] The structural genes of enzymes acetylcholine esterase (*Ace*) and of xanthine dehydrogenase (*rosy*) are included in the 12 complementation groups found from this segment of 315 kb.[27,149]

Results from the localization of polyadenylated RNA species derived from different stages of development of *D. melanogaster* in this region suggested that the largest chromomere units may contain several transcription units, but also nontranscribed DNA.[149] Further analyses of 43 distinct transcripts, the total length of them covering about 33% of the whole length of the 315-kb DNA segment have even better elucidated the distribution of transcription units in this area.[150] It seems that a proximal region of about 63 kb contains 18 different transcription units, while an adjacent, much longer region of about 153 kb apparently contains only 7 transcription units.[150] In cells of the salivary glands the number of detected transcripts corresponds to the number of chromomere bands (see Volume I, Chapter 12).

F. Developmental Genes in *Drosophila melanogaster*

The rapid progress in the cytogenetic and molecular analysis of genes and gene complexes affecting the embryogenesis of *Drosophila* has greatly increased the general interest in the genetic control of developmental processes also in other animals. Most well-known regions of the *Drosophila* genome controlling the segmentation are evidently the *Antennapedia* and *Bithorax* complexes locating in the 3R chromosome.

1. The Antennapedia Complex

Cytogenetic mapping of a number of mutants in the *Antennapedia (Antp)*[15] and in the adjacent *proboscipedia (pb)*[152] locus at the proximal part of 3R started a series of extensive analyses of this region[153-155] (cf. Chapter 16, Table 1 and EM map). The important role of the genes locating at the *Antennapedia complex* (ANT-C) in many developmental processes was demonstrated by further studies on functions of the loci in this region.[156,157] Isolation and analysis of genomic and cloned DNA including the ANT-C started molecular mapping of the region.[158] The distribution of *Antp* transcripts in *Drosophila* embryos was visualized,[159] and demonstrated clearly the high activity of *Antp* genes during the segmentation stages of development. Further analyses of even longer stretches of cloned DNA from the area of ANT-C have elucidated the distribution of mutations across the segment containing more than 100 kb of DNA.[160]

A relatively short repetitive DNA sequence, named as "homeo box", was found to be common to the genes *Antp, Ubx (Ultrabithorax)*, and *ftz (fushi tarazu)*.[161,162] These were all known to be active in the segmentation stage of the embryos of *D. melanogaster*. The *fushi tarazu* function appears as the reduction of the number of segments. Results and problems concerning the cytogenetic and molecular studies on ANT-C have been discussed more comprehensively in a recent review article of Scott.[163]

2. The Bithorax Complex

As pointed out by Lewis,[164] the *bithorax* region in the chromosome 3R of *D. melanogaster*, affecting the development of thoracal and abdominal segments, offers an excellent object to study the problems involved in regulation and function of a gene complex (cf. Volume I, Chapter 8, Figure 8). The genes belonging to the *bithorax complex* (BX-C) show interaction with other genes within the same complex[165] and also with some genes outside of the complex.[166-168] Studies on the structural and functional organization of the *bithorax* complex have been summarized e.g. in recent reviews.[169,170] The progress and results of the cloning and molecular analyses of the *bithorax* region and of related developmental loci, particularly those regulating the segmentation, have also been introduced in several other reviews.[171-175]

A developmental gene *Toll* affecting the determination of the dorsal-ventral axis in the early embryo has been localized into the subdivision 97D of the 3R chromosome, which is quite far from the complexes described above.[176]

3. Developmental Genes in Other Chromosomes

A homoeotic gene *sxc (super sex combs)*, which seems to regulate the functions of *bithorax complex* in *D. melanogaster*, has been localized into the right arm of the second chromosome at about the border area of divisions 41 and 42.[177]

An X chromosomal gene, *fs(1)K10*, which is involved in the determination of the dorsal-ventral polarity in the earliest phases of the embryogenesis of *D. melanogaster* maps to the region 2E2-F1[178] (cf. the EM map in Chapter 16, Figure 1). About 200 kb of DNA isolated and cloned from this region of the X chromosome has been analyzed by the chromosome walking method. Several developmental genes have been detected in the close vicinity of the *K10* locus.

A "maternal effect gene", *dorsal (dl)*, controlling the establishment of the dorsal-ventral polarity in early embryogenesis of *D. melanogaster*, has been localized into the subdivision 36C at the left arm of the second chromosome.[179] The developmental gene *engrailed (en)* has been localized into the subdivision 48A at the proximal part of the chromosome 2R. This gene affects the appearance of segment borders in the abdominal segments of *D. melanogaster*.[180,181] Like the other segmentation genes, the *engrailed* also contains a homeo box.[182] The homeo box is located close the 3′ end of an open reading frame of about 1.7 kb. The homeo box itself is composed of two segments separated by an insert of about 0.28

kb.[183] During the developmental stages, the *engrailed* locus encodes several transcripts ranging from 1.4 to 3.6 kb in size.

A gene complex designated as *decapentaplegic (dpp)* has been mapped genetically at 2—4.0, corresponding to the chromosomal site 22F1-3 in 2L. The locus controls the development of imaginal disks.[184] According to the EM map of the salivary gland 2L chromosome, the region 22F1-3 consists of four rather prominent bands, a doublet and two single bands. The DNA content, estimated on the basis the relative thicknesses of the bands, is about 130 to 140 kb per chromatid. Thus the size of the *dpp* complex resembles that of ANT-C and BX-C.

For recent reviews on developmental genes of *Drosophila* and other organisms, the reader is referred to the collection of papers published in the book of the *50th Symposia on Quantitative Biology at Cold Spring Harbor*, 1985 and summarized by Rubin.[185]

G. The 4th Chromosome

Cytogenetic analysis of the complementation loci included in the small chromosome 4 of *D. melanogaster* has been carried out mainly by Hochman and his collaborators by using both spontaneous and induced lethals.[186-190] On the basis of data from complementation tests on about 100 chemically induced and on more than 70 previously known, X-ray-induced and other lethals, as well as a few visible mutations, 43 genetic loci were determined in the fourth chromosome.[189,190] Twelve of the loci, nine of them vital ones, were localized into an area of a deficiency *Df(4)M* including about 11 to 15 bands. Seven or eight loci seem to be located in the area of a distal deficiency *Df(4)G* including 8 to 9 bands. The rest (24) of the detected complementation groups seem to be located in a region composed of about an equal number of bands in between the deficiencies mentioned above. Thus the total number of vital loci in the chromosome 4 is close to the number of bands drawn by C. B. Bridges in his map in 1935. In the EM map included in this book the bands of the division 102 have been numbered only. This map, compiled on the basis of electron micrographs of thin-sectioned 4th chromosomes, shows 40 bands for the division 102. However, ten of the prominent bands, although interpreted as single bands in the EM map, may appear as doublets in well-stretched chromosomes; thus the maximum number of bands in the last division 102 may be about 50. The thin sections used for the EM analysis of the 4th chromosome were not capable of showing more than about 10 bands from division 101 (cf. the maps in Chapter 16, Figure 1).

H. Genes with Multiple Chromosomal Sites

Autoradiography of ecdysone induced activity in polytene chromosomes has shown a label in several sites.[191] Accordingly, the hormonally regulated *Yps* genes coding for yolk protein precursors in *D. melanogaster* have been localized in several sites (8E, 9B, 12A, and 12D) in the X chromosome.[192] Analysis of DNA clones complementary to poly(A) mRNAs isolated from the follicle cells has revealed two distinct clusters of chorion protein genes. Genes coding for the two proteins designated as c36 and c38 are located in the area 7E11-7F2 in the X chromosome. The genes coding for the third presumptive chorion protein were localized at the chromosomal site 66D15 in 3L.[193] Later studies on chorion genes have shown two clusters of these genes in the chromosomal sites 7F1-2 in the X chromosome and 66D11-15 in 3L. A molecular analysis has revealed two tandemly transcribed copies of chorion genes in both of this clusters.[194]

Genes coding for α-tubulin have been localized in at least four bands.[195] One of the chromosomal sites seems to be 67C in 3L; three others 84B/C, 84D, and 85E are located in 3R chromosome. The two chromosomal sites detected for β-tubulin genes are 60A/B in 2R and 85D in 3R. *In situ* hybridization experiments on β-tubulin genes derived from genomic and cloned segments of DNA have demonstrated 3 chromosomal sites 56D and 60B in 2R and 85D in 3R.[196]

Another dispersive gene family are the actin genes. Tritium-labeled cRNA from the actin gene area has been shown to hybridize into five chromosomal sites, 5C, 57A, 79B, 87E, and 88F.[197] An isolated and cloned about 7.2-kb-long fragment (K1) of chromosomal DNA, complementary to actin genes, has been hybridized into the region 88F.[198] The actin gene localized in region 79B in 3L apparently codes for actin specific to the larval muscles.[199] According to present knowledge there are actin genes at sites 5C, 42A, 57A, 79B, 87E, and 88F in the salivary gland chromosomes of *D. melanogaster.*[200]

Apart from the gene loci introduced briefly above, there are some extensively studied groups of multiple site genes like the glue protein genes, heat shock protein genes, and the families of RNA genes.

1. The Glue Protein Genes

A group of genes found to be active in salivary gland cells in the multigene family is designated as *Sgs.* The members of this family are coding for salivary gland secretion proteins, i.e., the glue proteins.[201] The gene *Sgs-1* responsible for production of a main component of glue has been mapped to the chromosome 2L in region 25B.[202] The production of this protein seems to be correlated to the puffing activity observed in the area of subdivision 25B.[202] It has been demonstrated with polyacrylamide electrophoresis that the secretion of salivary gland cells in *D. melanogaster* larvae is composed of several protein components encoded by different genes. Production of one of the components Sgs-6 has appeared to be controlled by a gene located in the region bordered by bands 71C1-2 and 71E3-5 in the chromosome 3L. The presence of the Sgs-6 gene in the genome appears as an early puff at site of bands 71C3-4, which has been interpreted as a sign of possible activity of the Sgs-6 locus.[203]

Cytogenetic analysis of the Sgs-4 gene has allocated it in the region of small puffed bands 3C11 to 12 in the X chromosome.[204] EM studies of this region have shown that the area in question actually consists of 4 narrow dotted bands. The region is normally puffed in the salivary gland chromosomes of *D. melanogaster*, but not in the cells of certain mutants. It has been demonstrated that a segment locating adjacently at the distal side of the *Sgs-4* puff may control the activity of this saliva protein gene.[204] A small transposable element of only about 1.3 kb has been found to be inserted in some variant types of *Sgs-4* locus.[205] An insertion of this element between the remote area and the structural genes reduces the activity of the locus[205] (cf. Figure 4.)

A cluster of three *Sgs-3* glue protein genes at the site 68C in salivary gland 3L chromosome seems to be regulated by a gene designated according to the lethal l(1)npr-1. The presence of normal products of this control gene are needed to regress puffing at the *Sgs-3* site.[206]

2. Chromosomal Localization of Heat Shock Protein Genes

A number of reports concerning the heat-induced puffs in the salivary gland chromosomes of *D. melanogaster* have been already cited in Chapter 11 in Volume I.

Some of the heat shock loci have been subjected to further studies. For instance, chromosomal mapping of genes (*hsp*) coding for small heat shock proteins (22 to 27K) has shown that these genes or at least three of them are located in the area of the heat shock puff 67B in 3L.[207] Recently, several reports have also considered molecular studies on *hsp* DNA. Tandemly repeated transcription units have been detected in the molecular analyses of cloned segments of DNA derived from a major heat shock puff 87C.[208] Analysis of the sequence organization of DNA segments including genes for 70K heat shock protein has shown that the same sequence exists as seven repeats per haploid genome in *D. melanogaster.*[209] Copies of the same genes coding for the 70K hsp have been mapped to two different chromosomal sites, 87A and 87C. Refined mapping of these regions has shown that the sites of 70K hsp are located in the bands 87A7 and 87C1[210] (cf. Volume I, Chapter 11, Figure 4).

Genes of another major hsp (72K) have been localized in the subdivision 93D in the salivary gland 3R chromosome.[211] Allocation of 72K *hsp* genes has been refined by molecular cloning. Results from *in situ* hybridization of the cloned segment into salivary gland chromosomes have indicated that a cluster of *hsp* genes is located in the bands 93D 6-7.[212] Accordingly, the chromosomal site of the heat-inducible activity in the second chromosome of *D. hydei* has been recently allocated to the band 48B8 by using new cytogenetic methods.[213]

3. Chromosomal Sites of tRNA Genes

The very first RNA-DNA hybridization experiments already indicated that the RNAs representing the size class of tRNA molecules share homology with chromosomal DNA in numerous different regions of salivary gland chromosomes.[214] Purification and iodination of RNA samples made it possible to localize specific tRNA species quite exactly along the salivary gland chromosomes by means of *in situ* hybridization combined with the autoradiography.[215-218] Besides the major sites, most of the tRNA species seem to hybridize to several other regions in the polytene chromosomes.[219] For instance, the labeled transfer RNA^{Tyr} has been localized to eight different regions in the salivary gland chromosomes of *D. melanogaster*. This tRNA type seems to be transcribed by about 23 genes per haploid genome. The chromosomal sites of $tRNA^{Tyr}$ coding sequences are distributed to the following regions: four or five copies of $tRNA^{Tyr}$ genes are located in regions 22F and 85A, clusters of two or three copies are located in regions 28C, 41AB, 42A, 42E, and 56D.[220] Aside from tRNA genes, the genome of *D. melanogaster* seems to contain genes (U_{1-6}) coding for small nuclear RNAs.[221]

4. Localization of Genes for Ribosomal RNAs

The fact that rRNA is complementary to the nucleolar DNA was demonstrated in 1965.[222] Five years later the sites of ribosomal genes were shown in the polytene chromosomes of three species of *Diptera* by using *in situ* hybridization with autoradiography.[223] Again, 5 years after this the ribosomal genes of *D. melanogaster* were localized in the salivary gland chromosomes by using cloned segments of chromosomal DNA coding for the rRNA types 18S and 28S.[224] The arrangement of coding and noncoding sequences in the ribosomal genes of *D. melanogaster* has also been elucidated by electron microscopy of heteroduplex formation in the DNA level by using restriction fragments of cloned DNA complementary to ribosomal 18S and 28S RNA.[225]

Two major classes of repeating units, 17 and 11.4 kb, were found in 28S ribosomal genes of *D. melanogaster*. The size of the repeat unit seemed to be dependent on the presence or absence of an intervening sequence. A more accurate analysis of the organization of X- and Y-chromosomal repeat units of the ribosomal RNA genes and of their presence in the polytene nuclei of *Drosophila* have been carried out by using molecular methods.[226-232]

In the 28S rRNA genes of *D. virilis*, the longer interrupting sequence seems to be about 9.6 kb.[233] In accordance to the results obtained from *D. melanogaster*, in several sibling species also, the shorter repeat unit seems to be about 11 kb.[234] Comparison of the structure of rDNA in six sibling species of *D. melanogaster* has indicated that the nontranscribed spacer DNA between actual coding sequences is, at least in four species, more conserved than the structure of genes.[235] Closer analysis of spacer sequences from other species of *Drosophilia* has revealed about 20 to 30% homology between the species *D. virilis* and *D. hydei* both belonging to the subgenus *Drosophila*. On the other hand, very little if any homology has been found in spacers between *D. virilis* and *D. melanogaster* (belonging to a different subgenus).[235]

Accordingly, the spacer units of rRNA genes of the species *D. hydei* and *D. melanogaster* are very different from each other. However, in all of these species the spacers seem to contain repetitive units of about 0.25 kb elements.[235,236] Certain conservation has been found in the chromosomal location of nucleolus organizer regions in species belonging to the repleta group of the subgenus *Drosophila*. Distribution of the radioactive label in polytene chromosomes after hybridization with rRNA suggests that clusters of ribosomal RNA genes in *D. neohydei* exist besides the X chromosome also in some autosomes.[237]

I. Chromosomal Localization of Mobile DNA Elements

The different types of mobile DNA elements found in the cells of *D. melanogaster* have been already introduced in Volume I, Chapter 7, Section III (see also Table 1).

Among the first nomadic chromosome segments mapped in the salivary gland chromosomes of *D. melanogaster* was the large transposable element (*TE*) carrying a marker gene w^a. Starting from early findings of the peculiar appearance of white apricot eye color caused by a gene localized to the second chromosome, it soon became evident that the segment including this gene was nomadic, i.e. the gene was apparently attached to a transposable element.[238-240] Due to the frequent jumping of the element, all of the new chromosomal positions of this element could not be mapped. For the genetic and chromosomal sites of *TE* elements, see the reports and reviews of Ising et al[52,70,240] (see also Section II.A.4). For the genetic and chromosomal sites of *TE* elements, see the reports and reviews of Ising et al.[52,70,240] (see also Section II.A.4). Since the *TE* transposable elements carry smaller foldback (FB) DNA segments which are resposible for the mobility of whole units, the *TE* type of elements are included in the class of transposable DNA elements with long inverted repeats.[241]

A catalog of transposable elements found in the genome of *D. melanogaster* is presented in Table 1. Chromosomal hybridization sites of cloned dispersed sequences have been listed in several reports, e.g. in References 242 to 248. Comprehensive catalogs of chromosomal sites of cloned DNA segments in the salivary gland chromosomes of *D. melanogaster* have been published in *DIS (Drosophila Information Service)* by Merriam et al.,[249,250] (cf. the Table 1 in Chapter 16).

Table 1
TRANSPOSABLE ELEMENTS IN GENOME OF *DROSOPHILA MELANOGASTER*

Element	Length (in kb)	NRS	TIR (bp)	TSD (bp)	CN	LPPC
Sancho 2	2.6	7	—	—	30	1B1-2(Hw)
Jockey	2.8	6	—	—	1—?	7B3-4 (ct)
P	2.9	9	31	8	0—50	3C2(w)
hobo	3.0	11	12	8	20—50	3C11-12(sgs-4)
Doc	4.3	6	—	—	1—?	89E1-4(bx)
1731	4.4	?	350	?	?	?
Sancho 1	4.5	8	—	—	50	1B3(sc)
Kermit	4.8	5(7)	—	—	30	87E1-6
copia	5.1	5	13/17	5	60	3C2(w); 89E1-4(bx); + about 40 other positions; see Reference 247
I	5.37	8	—	12—14	0—15	89E1-4(bx)
mdg3	5.4	9	4/5	4	15	About 25 mapped positions; see Reference 247
NEB	5.5	5	5	—	?	TE98rst
3S18	6.5	17	5	—	15	3A1(gt); 3C2(w); 89E1-4(bx)
297	7.0	9	4/5	4	30	?
Delta	7.0	6	—	—	?	89E(tuh-3)
Calypso	7.2	11	—	—	10—20	87D1-2(ry)
Harvey	7.2	5	—	—	?	89E1-4(bx)
BEL	7.3	7	—	—	25	3C2(w)
HMS Beagle	7.3	5	6/7	4	50	44D(cp3)
mdg1	7.3	8	13/16	4	25	About 35 mapped positions; see Reference 247
mdg4/gypsy	7.3	10	4/5	4	10	1A5-8(y^2); 1B1-2(Hw); 1B1-3(sc); 7B3-4(ct); 15F1-5(f); 89E(bx,bxd); su(Hw) + about 7 other positions; see Reference 247
17.6	7.4	8	—	4	40	39DE(his-family)
412	7.6	9	5/6	4	40	10A1(v); 89E1-4(bx)
BS	8.0	7	2500	—	15	1B1-2(Hw)
B104/roo	8.7	15	3	5	80	3C2(w); 3C7(N); 3C11-(sgs4)10A1(v); 84B1-2(Antp); 89E1-4(bx)
Springer	8.8	7	6/10	6	6	88C-E(lfm3)
Elements with variable size						
F	variable (4.7)	5	—	8—22	50	3C2(w^i); 101F
FB	variable	About 50	About 1 kb	9	30	3C2(w^c,w^{DZL})
G	variable (4kb)	7	—	9	10—20	20DF(rDNA)

Table 1 (continued)
TRANSPOSABLE ELEMENTS IN GENOME OF *DROSOPHILA MELANOGASTER*

Note: Transposable elements in the genome of *D. melanogaster*. The lengths of elements given approximately in kilobase pairs (kb). Explanations for the columns 3 to 7. NRS = number of restriction sites identified in the area of element, TIR = length of the terminal inverted repeat in number of base pairs (bp), TSD = length of the target site duplication by insertion of element in bp, CN = approximate copy number per genome, LPPC = localized chromosomal positions in the polytene chromosomes of *D. melanogaster*. For more details concerning, for instance, the restriction maps, lists of effective and of noneffective restriction enzymes in each case, as well as the sequenced ends of elements, see the appendix of the review of Finnegan and Fawcett.[241]

Compiled according to Finnegan, P. J. and Fawcett, D. H., in the *Oxford Surveys on Eukaryotic Genes*, Vol. 3, 1, 1986.

REFERENCES

1. **Lefevre, G., Jr. and Wilkins, M. D.,** Cytogenetic studies on the *white* locus in *Drosophila melanogaster, Genetics,* 53, 175, 1966.
2. **Sorsa, V., Green, M. M., and Beermann, W.,** Cytogenetic fine structure and chromosomal localization of the *white* gene in *Drosophila melanogaster, Nature, New Biol.,* 245, 43, 1973.
3. **O'Hare, K., Murphy, C., Levis, R., and Rubin, G. M.,** DNA sequence of the *white* locus of *Drosophila melanogaster, J. Mol. Biol.,* 180, 437, 1984.
4. **Zhimulev, I. F., Belaeva, E. S., Semeshin, V. F., Pokholkova, G. V., and Grafodatskaya, V. E.,** On the structural and functional organization of polytene chromosomes, *Proc. 14th Int. Congr. Genet.,* 3(2), 271, 1980.
5. **Zhimulev, I. F., Semeshin, V. F., and Belyaeva, E. S.,** Fine cytogenetical analysis of the band 10A1-2 and the adjoining regions in the *Drosophila melanogaster* X chromosome. I. Cytology of the region and mapping of chromosome rearrangements, *Chromosoma,* 82, 9, 1981.
6. **Judd, B. H., Shen, M. W., and Kaufman, T. C.,** The anatomy and function of a segment of the X chromosome of *Drosophila melanogaster, Genetics,* 71, 139, 1972.
7. **Kaufman, T. C., Shannon, M. P., Shen, M. W., and Judd, B. H.,** A revision of the cytology and ontogeny of the several deficiencies in the 3A1-3C6 region of the X chromosome of *Drosophila melanogaster, Genetics,* 79, 265, 1975.
8. **Hochman, B.,** Analysis of whole chromosome in *Drosophila melanogaster, Cold Spring Harbor Symp. Quant. Biol.,* 38, 581, 1974.
9. **Fujita, S.,** Chromosomal organization as the genetic basis of cytodifferentiation in multicellular organisms, *Nature,* 206, 742, 1965.
10. **Crick, F.,** General model for the chromosomes of higher organisms, *Nature,* 234, 25, 1971.
11. **Speiser, C.,** Eine Hypothese über die funktionelle Organisation der Chromosomen höherer Organismen, *Theor. Appl. Genet.,* 44, 97, 1974.
12. **Gersch, E. S.,** Sites of gene activity and of inactive genes in polytene chromosomes of *Diptera, J. Theor. Biol.,* 50, 413, 1975.
13. **Zhimulev, I. F. and Belyaeva, E. S.,** Proposals to the problem of structural and functional organization of polytene chromosomes, *Theor. Appl. Genet.,* 45, 335, 1975.
14. **Haldane, J. B. S.,** The combination of linkage values, and the calculation of distances between the loci of linked factors, *J. Genet.,* 8, 299, 1919.
15. **Emerson, R. A. and Rhoades, M.,** Relation of chromatid crossing-over to the upper limit of recombination percentages, *Am. Nat.,* 67, 374, 1933.
16. **Mather, K.,** Crossing-over, *Biol. Rev. Cambridge Philos. Soc.,* 13, 252, 1938.
17. **Judd, B. H.,** Mapping the functional organization of eukaryotic chromosomes, in *Cell Biology,* Vol. 2, Prescott, D. M. and Goldstein, L., Eds., Academic Press, New York, 1979, 223.

18. **Jackson, D., Symons, R., and Berg, P.,** Biochemical methods for inserting new genetic information into DNA of Simian Virus 40: *SV40* DNA molecules containing *Lambda* phage genes and the galactose operon of *Escherichia coli, Proc. Natl. Acad. Sci. U.S.A.,* 69, 2904, 1972.
19. **Lobban, P. E. and Kaiser, A. D.,** Enzymatic end-to-end joining of DNA molecules, *J. Mol. Biol.,* 78, 453, 1973.
20. **Thomas, M., Cameron, J. R., and Davis, R. W.,** Viable molecular hybrids of bacteriophage lambda and eukaryotic DNA, *Proc. Natl. Acad. Sci. U.S.A.,* 71, 4579, 1974.
21. **Murray, N. E. and Murray, K.,** Manipulation of restriction targets in phage *lambda* to form receptor chromosomes for DNA fragments, *Nature,* 251, 476, 1974.
22. **Wensink, P. C., Finnegan, D. J., Donelson, J. E., and Hogness, D. S.,** A system for mapping DNA sequences in the chromosomes of *Drosophila melanogaster, Cell,* 3, 315, 1974.
23. **Gall, J. G. and Pardue, M. L.,** Formation and detection of RNA-DNA hybrid molecules in cytological preparations, *Proc. Natl. Acad. Sci. U.S.A.,* 63, 378, 1969.
24. **Cohen, E. H. and Bowman, S. C.,** Detection and location of three simple sequence DNAs in polytene chromosomes of *virilis* group species of *Drosophila, Chromosoma,* 73, 327, 1979.
25. **Hennig, W.,** Highly repetitive DNA sequences in the genome of *Drosophila hydei.* II. Occurrence in polytene tissues, *J. Mol. Biol.,* 71, 419, 1972.
26. **Maxam, A. M. and Gilbert, W.,** A new method for sequencing DNA, *Proc. Natl. Acad. Sci. U.S.A.,* 74, 560, 1977.
27. **Spierer, P., Spierer, A., Bender, W., and Hogness, D. S.,** Molecular mapping of genetic and chromomeric units in *Drosophila melanogaster, J. Mol. Biol.,* 168, 17, 1983.
28. **Bender, W., Spierer, P., and Hogness, D. S.,** Chromosome walking and jumping to isolate DNA from the *Ace* and *rosy* loci and the *bithorax* complex in *Drosophila melanogaster, J. Mol. Biol.,* 168, 17, 1983.
29. **Grundstein, M. and Hogness, D. S.,** Colony hybridization: a method for the isolation of cloned DNA's that contain a specific gene, *Proc. Natl. Acad. Sci. U.S.A.,* 72, 3961, 1975.
30. **White, R. L. and Hogness, D. S.,** R-loop mapping of the 18S and 28S sequences in the long and short repeating units of *Drosophila melanogaster* rDNA, *Cell,* 10, 177, 1977.
31. **Bridges, C. B.,** A revised map of the salivary gland X-chromosome of *Drosophila melanogaster, J. Hered.,* 29, 11, 1938.
32. **Lefevre, G.,** The distribution of randomly recovered X-ray-induced sex-linked genetic effects in *Drosophila melanogaster, Genetics,* 99, 461, 1981.
33. **Shannon, M. P., Kaufman, T., Shen, M. W., and Judd, B. H.** Lethality patterns and morphology of selected lethal and semi-lethal mutations in the *zeste-white* region of *Drosophila melanogaster, Genetics,* 72, 615, 1972.
34. **Hazelrigg, T.,** The *Drosophila white* gene: a molecular update, *Trends in Genetics,* 3, 43, 1987.
35. **Sorsa, V.,** Electron microscopic localization and ultrastructure of certain gene loci in salivary gland chromosomes of *Drosophila melanogaster,* in *Specific Eukaryotic Genes, Alfred Benzon Symp.,* 13, 55, 1979.
36. **Sorsa, V.,** An attempt to estimate DNA content and distribution in the *zeste-white* region of the X chromosome of *Drosophila melanogaster, Biol. Zbl.,* 101, 81, 1982.
37. **Sorsa, V.,** Electron microscopic map for the salivary gland X chromosome of *Drosophila melanogaster* divisions 1-5, in *Advances in Genetics, Development and Evolution of Drosophila,* Lakovaara, S., Ed., Plenum Press, New York, 1982, 23.
38. **Young, M. W. and Judd, B. H.,** Nonessential sequences, genes, and the polytene chromosome bands of *Drosophila melanogaster, Genetics,* 88, 723, 1978.
39. **Jack, J. W. and Judd, B. H.,** Allelic pairing and gene regulation: a model for the *zeste-white* interaction in *Drosophila melanogaster, Proc. Natl. Acad. Sci. U.S.A.,* 76, 1368, 1979.
40. **Lifschytz, E. and Green, M. M.,** The *zeste-white* interaction: induction and genetic analysis of a novel class of zeste alleles, *The EMBO J.,* 3, 999, 1984.
41. **Gubb, D., Roote, J., McGill, S., Shelton, M., and Ashburner, M.,** Interactions between white genes carried by a large transposing element and the *zeste* allele in *Drosophila melanogaster, Genetics,* 112, 551, 1986.
42. **Green, M. M.,** The genetics of a mutable gene at the *white* locus of *Drosophila melanogaster, Genetics,* 56, 467, 1967.
43. **McClintock, B.,** Chromosome organization and genic expression, *Cold Spring Harbor Symp. Quant. Biol.,* 16, 13, 1951.
44. **McClintock, B.,** The control of gene action in maize, *Brookhaven Symp. Biol.,* 18, 162, 1965.
45. **Green, M. M.,** Mapping a *Drosophila melanogaster* "controlling element" by interallelic crossing over, *Genetics,* 61, 423, 1969.
46. **Lefevre, G., Jr. and Green, M. M.,** Genetic duplication in the *white-split* interval of the X chromosome in *Drosophila melanogaster, Chromosoma,* 36, 391, 1972.
47. **Bridges, C. B.,** Salivary chromosome maps with a key to the banding of the chromosomes of *Drosophila melanogaster, J. Hered.,* 26, 60, 1935.

48. **Judd, B. H.,** Genetic units of *Drosophila* complex loci, in *The Genetics and Biology of Drosophila,* Vol. 1b, Ashburner, M. and Novitski, E., Eds., Academic Press, New York, 1976, 767.
49. **Sorsa, V.,** Organization of chromomeres, *Cold Spring Harbor Symp. Quant. Biol.,* 38, 601, 1974.
50. **Sorsa, V.,** Electron microscopic mapping and ultrastructure of *Drosophila* polytene chromosomes, in *Insect Ultrastructure,* Vol. 2, King, R. C. and Akai, H., Eds., Plenum Press, New York, 1984, 75.
51. **Pirrotta, V. and Brõckl, C.,** Transcription of the *Drosophila white* locus and some of its mutants, *The EMBO J.,* 3, 563, 1984.
52. **Ising, G. and Ramel, C.,** The behavior of a transposing element in *Drosophila melanogaster,* in *The Genetics and Biology of Drosophila,* Vol. 1b, Ashburner, M. and Novitski, E., Eds., Academic Press, New York, 1976, 947.
53. **Gehring, W. J. and Paro, R.,** Isolation of a hybrid plasmid with homologous sequences to a transposing element of *Drosophila melanogaster, Cell,* 19, 897, 1980.
54. **Goldberg, M. L., Paro, R., and Gehring, W. J.,** Molecular cloning of the *white* locus region of *Drosophila melanogaster* using a transposable element, *The EMBO J.,* 1, 93, 1982.
55. **Paro, R., Goldberg, M. L., and Gehring, W. J.,** Molecular analysis of large transposable elements carrying the *white* locus of *Drosophila melanogaster, The EMBO J.,* 2, 853, 1983.
56. **Bingham, P. M., Levis, R., and Rubin, G. M.,** Cloning of DNA sequences from the *white* locus of *Drosophila melanogaster* by a novel and general method, *Cell,* 25, 693, 1981.
57. **Bingham, P. M. and Judd, B. H.,** A copy of the *copia* transposable element is very tightly linked to the w^a allele at the *white* locus of *D. melanogaster, Cell,* 25, 705, 1981.
58. **Levis, R., Bingham, P. M., and Rubin, G. M.,** Physical map of the *white* locus of *Drosophila melanogaster, Proc. Natl. Acad. Sci. U.S.A.,* 79, 564, 1982.
59. **Karess, R. E. and Rubin, G. M.,** A small tandem duplication is responsible for the unstable *white-ivory* mutation in *Drosophila, Cell,* 30, 63, 1982.
60. **Collins, M. and Rubin, G. M.,** Structure of the *Drosophila* mutable allele, *white-crimson,* and its *white-ivory* and wild-type derivatives, *Cell,* 30, 71, 1982.
61. **Zachar, Z. and Bingham, P. M.,** Regulation of *white* locus expression: the structure of mutant alleles at the *white* locus of *Drosophila melanogaster, Cell,* 30, 529, 1982.
62. **Bingham, P. M.,** A novel dominant mutant allele at the *white* locus of *Drosophila melanogaster* is mutable, *Cold Spring Harbor Symp. Quant. Biol.,* 45, 519, 1981.
63. **Levis, R. and Rubin, G. M.,** The unstable w^{DZL} mutation of *Drosophila* is caused by a 13 kilobase insertion that is imprecisely excised in phenotypic revertants, *Cell,* 30, 543, 1982.
64. **Levis, R., Collins, M., and Rubin, G. M.,** FB elements are the common basis for the instability of the w^{DZL} and w^c *Drosophila* mutations, *Cell,* 30, 551, 1982.
65. **Pirrotta, V., Hadfield, C., and Pretorius, G. H. J.,** Microdissection and cloning of the *white* locus and the 3B1-3C2 region of the *Drosophila* X chromosome, *The EMBO J.,* 2, 927, 1983.
66. **O'Hare, K., Levis, R., and Rubin, G. M.,** Transcription of the *white* locus in *Drosophila melanogaster, Proc. Natl. Acad. Sci. U.S.A.,* 80, 6917, 1983.
67. **O'Hare, K. and Rubin, G. M.,** Structures of P transposable elements and their sites of insertion and excision in the *Drosophila melanogaster* genome, *Cell,* 34, 25, 1983.
68. **Gehring, W. J., Klemenz, R., Weber, U., and Kloter, U.,** Functional analysis of the *white*$^+$ gene of *Drosophila* by P-factor-mediated transformation, *The EMBO J.,* 3, 2077, 1984.
69. **Goldberg, M. L., Sheen, J.-Y., Gehring, W., and Green, M. M.,** Unequal crossing-over associated with asymmetrical synapsis between nomadic elements in the *Drosophila melanogaster* genome, *Proc. Natl. Acad. Sci. U.S.A.,* 80, 5017, 1983.
70. **Ising, G. and Block, K.,** A transposon as a cytogenetic marker in *Drosophila melanogaster, Mol. Gen. Genet.,* 196, 6, 1984.
71. **Gubb, D., Shelton, M., Roote, J., McGill, S., and Ashburner, M.,** The genetic analysis of a large transposing element of *Drosophila melanogaster.* The insertion of a w^+ rst^+ TE into the *ck* locus, *Chromosoma,* 91, 54, 1984.
72. **Gersch, E. S.,** A new locus in the *white-Notch* region of the *Drosophila melanogaster* X chromosome, *Genetics,* 51, 477, 1965.
73. **Welshons, W. J.,** The analysis of a gene in *Drosophila, Science,* 150, 1122, 1965.
74. **Welshons, W. J.,** Genetic basis for two types of recessive lethality at the *Notch* locus of *Drosophila, Genetics,* 68, 259, 1971.
75. **Welshons, W. J.,** The cytogenetic analysis of a fractured gene in *Drosophila, Genetics,* 76, 775, 1974.
76. **Welshons, W. J. and von Halle, E. S.,** Pseudoallelism at the *Notch* locus in *Drosophila, Genetics,* 47, 743, 1962.
77. **Welshons, W. J. and Keppy, D. O.,** Intragenic deletions and salivary band relationships in *Drosophila, Genetics,* 80, 143, 1975.
78. **Keppy, D. O. and Welshons, W. J.,** The cytogenetics of a recessive visible mutant associated with a deficiency adjacent to the *Notch* locus in *Drosophila melanogaster, Genetics,* 85, 497, 1977.

79. **Keppy, D. O. and Welshons, W. J.**, The synthesis of compound bands in *Drosophila melanogaster* salivary gland chromosomes, *Chromosoma,* 76, 191, 1980.
80. **Welshons, W. J. and Keppy, D. O.**, The recombinational analysis of aberrations and the position of the *Notch* locus on the polytene chromosome of *Drosophila, Mol. Gen. Genet.*, 181, 319, 1981.
81. **Foster, G. G.**, Negative complementation at the *Notch* locus of *Drosophila melanogaster, Genetics,* 81, 99, 1975.
82. **Thōrig, G. E. W. and Scharloo, W.**, The action of the *Notch* locus in *Drosophila melanogaster.* III. Biochemical effects of recessive visible mutations on mitochondrial enzymes, *Genetica,* 57, 219, 1982.
83. **Artavanis-Tsakonas, S., Muskavitch, M. A. T., and Yedvobnick, B.**, Molecular cloning of *Notch,* a locus affecting neurogenesis in *Drosophila melanogaster, Proc. Natl. Acad. Sci. U.S.A.*, 80, 1977, 1983.
84. **Kidd, S., Lockett, T. J., and Young, M. W.**, The *Notch* locus of *Drosophila melanogaster, Cell,* 34, 421, 1983.
85. **Wharton, K., Johansen, K. M., Xu, T., and Artavanis-Tsakonas, S.**, Nucleotide sequence from the neurogenic locus *Notch* implies a gene product that shares homology with proteins containing EGF-like repeats, *Cell,* 43, 567, 1985.
86. **Grimwade, B. G., Muskavitch, M. A. T., Welshons, W. J., Yedvobnick, B., and Artavanis-Tsakonas, S.**, The molecular genetics of the *Notch* locus in *Drosophila melanogaster, Dev. Biol.*, 107, 503, 1985.
87. **Wharton, K. A., Yedvobnick, B., Finnerty, V. G., and Artavanis-Tsakonas, S.**, opa: a novel family of transcribed repeats shared by the *Notch* locus and other developmentally regulated loci in *D. melanogaster, Cell,* 40, 55, 1985.
88. **Welshons, W. J. and Welshons, H. J.**, Enhancement and suppression of a euchromatic position effect at *Notch* in *Drosophila, Genetics,* 113, 1986.
89. **Belyaeva, E. S., Aizenzon, M. G., Semeshin, V. F., Kiss, I. I., Koczka, K., Baricheva, E. M., Gorelova, T. D., and Zhimulev, I. F.**, Cytogenetic analysis of the 2B3-4-2B11 region of the X-chromosome of *Drosophila melanogaster.* I. Cytology of the region and mutant complementation groups, *Chromosoma,* 81, 281, 1980.
90. **Zhimulev, I. F., Vlassova, I. E., and Belyaeva, E. S.**, Cytogenetic analysis of the 2B3-4-2B11 region of the X chromosome of *Drosophila melanogaster.* III. *Chromosoma,* 85, 659, 1982.
91. **Konopka, R. J. and Benzer, S.**, Clock mutants of *Drosophila melanogaster, Proc. Natl. Acad. Sci. U.S.A.*, 68, 2112, 1971.
92. **Reddy, P., Zehring, W. A., Wheeler, D. A., Pirrotta, V., Hadfield, C., Hall, J. C., and Rosbash, M.**, Molecular analysis of the *period* locus in *Drosophila melanogaster* and identification of a transcript involved in biological rhythms, *Cell,* 38, 701, 1984.
93. **Zehring, W. A., Wheeler, D. A., Reddy, P., Konopka, R. J., Kyriacou, C. P., Rosbash, M., and Hall, J. C.**, P-element transformation with *period* locus DNA restores rhythmicity to mutant, Arrhythmic *Drosophila melanogaster, Cell,* 39, 369, 1984.
94. **Banga, S. S., Bloomquist, B. T., Brodberg, R. K., Pye, Q. N., Mason, J. M., Boyd, J. B., and Pak, W. L.**, Cytogenetic characterization of the 4BC region on the X chromosome of *Drosophila melanogaster:* localization of the *mei-9, norpA* and *omb* genes, *Chromosoma,* 93, 341, 1986.
95. **Jack, J. W.**, Molecular organization of the *cut* locus of *Drosophila melanogaster, Cell,* 42, 869, 1985.
96. **Zhimulev, I. F., Pokholkova, G. V., Bgatov, A. V., Semeshin, V. F., and Belyaeva, E. S.**, Fine cytogenetical analysis of the band 10A1-2 and the adjoining regions in the *Drosophila melanogaster* X chromosome. II. Genetical analysis, *Chromosoma,* 82, 25, 1981.
97. **Zhimulev, I. F. and Belyaeva, E. S.**, On the information content of polytene chromosome chromomeres, *Biol. Zbl.*, 104, 633, 1985.
98. **Freund, J. N., Vergis, W., Schedl, P., and Jarry, B. P.**, Molecular organization of the *rudimentary* gene of *Drosophila melanogaster, J. Mol. Biol.*, 189, 25, 1986.
99. **Lifschytz, E. and Falk, R.**, Fine structure analysis of a chromosome segment in *Drosophila melanogaster.* Analysis of X-ray induced lethals, *Mut. Res.*, 6, 235, 1968.
100. **Lifschytz, E.**, Fine structure analysis of the chromosome. Recombinational patterns at the base of the X chromosome of *Drosophila melanogaster, Mut. Res.*, 13, 35, 1971.
101. **Schalet, A. and Singer, K.**, A revised map of genes in the proximal region of the X chromosome of *Drosophila melanogaster, Drosophila Inf. Serv.*, 46, 131, 1971.
102. **Schalet, A. and Lefevre, G., Jr.**, The localization of ordinary sex-linked genes in section 20 of the polytene X chromosome of *Drosophila melanogaster, Chromosoma,* 44, 183, 1973.
103. **Lifschytz, E.**, Fine-structure analysis and genetic organization at the base of the X chromosome in *Drosophila melanogaster, Genetics,* 88, 457, 1978.
104. **Reuter, G. and Szidonya, J.**, Cytogenetic analysis of variegation suppressors and a dominant temperature-sensitive lethal in region 23-26 of chromosome 2L in *Drosophila melanogaster, Chromosoma,* 88, 277, 1983.

105. **Woodruff, R. C. and Ashburner, M.,** The genetics of a small autosomal region of *Drosophila melanogaster* containing the structural gene for alcohol dehydrogenase. I. Characterization of deficiencies and mapping of *Adh* and visible mutations, *Genetics,* 92, 117, 1979.
106. **Goldberg, D. A., Posakony, J. W., and Maniatis, T.,** Correct developmental expression of a cloned alcohol dehydrogenase gene transduced into the *Drosophila* germ line, *Cell,* 34, 59, 1983.
107. **Chia, W., Karp, R., McGill, S., and Ashburner, M.,** Molecular analysis of the *Adh* region of the genome of *Drosophila melanogaster, J. Mol. Biol.,* 186, 689, 1985.
108. **Rubin, G. M. and Spradling, A. C.,** Genetic transformation of *Drosophila* with transposable element vectors, *Science,* 218, 348, 1982.
109. **Spradling, A. C. and Rubin, G. M.,** Transposition of cloned P elements into germline chromosomes, *Science,* 218, 341, 1982.
110. **Scholnick, S. B., Morgan, B. A., and Hirsh, J.,** The cloned dopa decarboxylase gene is developmentally regulated when reintegrated into the *Drosophila* genome, *Cell,* 34, 37, 1983.
111. **Wright, T. R. F., Beermann, W., Marsh, J. L., Bishop, C. P., Steward, R., Black, B. C., Tomsett, A. D., and Wright, E. Y.,** The genetics of dopa decarboxylase in *Drosophila melanogaster.* IV. The genetics and cytology of the 37B10-37D1 region, *Chromosoma,* 83, 45, 1981.
112. **Rozek, C. E. and Davidson, N.,** *Drosophila* has one myosin heavy-chain gene with three developmentally regulated transcripts, *Cell,* 32, 23, 1983.
113. **Lifton, R. P., Goldberg, M. L., Karp, R. W., and Hogness, D. S.,** The organization of the histone genes in *Drosophila melanogaster* Functional and evolutionary implications, *Cold Spring Harbor Symp. Quant. Biol.,* 42, 1047, 1977.
114. **Saigo, K., Millstein, L., and Thomas, C. A., Jr.,** The organization of *Drosophila melanogaster* histone genes, *Cold Spring Harbor Symp. Quant. Biol.,* 45, 815, 1981.
115. **Pardue, M. L., Kedes, L. H., Weinberg, E. S., and Birnstiel, M. L.,** Localization of sequences coding for histone messenger RNA in the chromosomes of *Drosophila melanogaster, Chromosoma,* 63, 135, 1977.
116. **Khesin, R. B., and Leibovitch, B. A.,** Influence of deficiency of the histone gene-containing 38B-40 region on X-chromosome template activity and the *white* gene position effect variegation in *Drosophila melanogaster, Mol. Gen. Genet.,* 162, 323, 1978.
117. **Siegel, J. G.,** Genetic characterization of the region of the *Drosophila* genome known to include the histone structural gene sequences, *Genetics,* 98, 505, 1981.
118. **Moore, G. D., Sinclair, D. A., and Grigliatti, T. A.,** Histone gene multiplicity and position effect variegation in *Drosophila melanogaster, Genetics,* 105, 327, 1984.
119. **Bridges, C. B. and Bridges, P. N.,** A revised map of the right limb of the second chromosome of *Drosophila melanogaster, J. Hered.,* 30, 475, 1939.
120. **Wimber, D. E. and Steffensen, D. M.,** Localization of 5S RNA genes on *Drosophila* chromosomes by RNA-DNA hybridization, *Science,* 170, 639, 1970.
121. **Tartof, K. D. and Perry, R. P.,** The 5S RNA genes of *Drosophila melanogaster, J. Mol. Biol.,* 51, 171, 1970.
122. **Bencze, J. L., Brasch, K., and White, B. N.,** The location of 5S RNA genes in lampbrush polytene chromosomes from *Drosophila, Exp. Cell Res.,* 120, 365, 1979.
123. **Procunier, J. D. and Dunn, R. J.,** Genetic and molecular organization of the 5S locus and mutants in *D. melanogaster, Cell,* 15, 1087, 1978.
124. **Artavanis-Tsakonas, S., Schedl, P., Tschudi, C., Pirrotta, V., Steward, R., and Gehring, W. J.,** The 5S RNA genes of *Drosophila melanogaster, Cell,* 12, 1057, 1977.
125. **Alonso, C. and Berendes, H. D.,** The location of 5S(ribosomal) RNA genes in *Drosophila hydei, Chromosoma,* 51, 347, 1975.
126. **Wimber, D. E. and Wimber, D. R.,** Sites of the 5S ribosomal genes in *Drosophila.* I. The multiple clusters in the *virilis* group, *Genetics,* 86, 133, 1977.
127. **Cohen, M., Jr.,** Evolution of 5S ribosomal RNA genes in the chromosomes of the *virilis* group of *Drosophila, Chromosoma,* 55, 359, 1976.
128. **Cohen, M., Jr.,** Ectopic pairing and evolution of 5S ribosomal RNA genes in the chromosomes of *Drosophila funebris, Chromosoma,* 55, 349, 1976.
129. **Hershey, N. D., Conrad, S. E., Sodja, A., Yen, P. H., Cohen, M., Jr., and Davidson, N.,** The sequence arrangement of *Drosophila melanogaster* 5S DNA cloned in recombinant plasmids, *Cell,* 11, 585, 1977.
130. **Wimber, D. E. and Steffensen, D. M.,** Localization of gene function, *Ann. Rev. Genet.,* 7, 205, 1973.
131. **Snyder, M., Hirsh, J., and Davidson, N.,** The cuticle genes of *Drosophila*: a developmentally regulated gene cluster, *Cell,* 25, 165, 1981.
132. **Snyder, M., Hunkapiller, M., Yen, D., Silvert, D., Fristrom, J., and Davidson, N.,** Cuticle protein genes of *Drosophila:* structure, organization and evolution of four clustered genes, *Cell,* 29, 1027, 1982.
133. **Izquierdo, M., Arribas, C., and Alonso, C.,** Isolation of a structural gene mapping to subregions 63F of *Drosophila melanogaster* and 90B of *D. hydei* polytene chromosomes, *Chromosoma,* 83, 353, 1981.

134. **Mōritz, Th., Edstrōm, J. E., and Pongs, O.,** Cloning of a gene localized and expressed at the ecdysteroid regulated puff 74EF in salivary glands of *Drosophila* larvae, *The EMBO J.*, 3, 289, 1984.
135. **Akam, M. E., Roberts, D. B., Richards, G. P., and Ashburner, M.,** *Drosophila:* the genetics of two major larval proteins, *Cell,* 13, 215, 1978.
136. **Holmgren, R.,** Cloning sequences from the *hairy* gene of *Drosophila, The EMBO J.*, 3, 569, 1984.
137. **Bridges, P. N.,** A revision of the salivary gland 3R-chromosome map of *Drosophila melanogaster, J. Hered.*, 32, 299, 1941.
138. **O'Tousa, J. E., Baehr, W., Martin, R. L., Hirsh, J., Pak, W. L., and Applebury, M. L.,** The *Drosophila ninaE* gene encodes an opsin, *Cell,* 40, 839, 1985.
139. **Cowman, A. F., Zuker, C. S., and Rubin, G. M.,** An opsin gene expressed in only one photoreceptor cell type of the *Drosophila* eye, *Cell,* 44, 705, 1986.
140. **Vincent, A., O'Connell, P., Gray, M. R., and Rosbash, M.,** *Drosophila* maternal and embryo mRNAs transcribed from a single transcription unit use alternate combinations of exons, *The EMBO J.*, 3, 1003, 1984.
141. **Morrison, W. J. and MacIntyre, R. J.,** Cytogenetic localization of the acid phosphatase-1 gene in *Drosophila melanogaster, Genetics,* 88, 487, 1978.
142. **Chovnick, A., Gelbart, W., and McCarron, M.,** Organization of the *rosy* locus in *Drosophila melanogaster, Cell,* 11, 1, 1977.
143. **Chovnick, A., McCarron, M., Hilliker, A., O'Donnel, J., Gelbart, W., and Clark, S.,** Organization of a gene in *Drosophila:* a progress report, *Stadler Symp.*, 10, 9, 1978.
144. **Gelbart, W. M. and Chovnick, A.,** Spontaneous unequal exchange in *rosy* region of *Drosophila melanogaster, Genetics,* 92, 849, 1979.
145. **Gausz, J., Bencze, G., Gyuorkovicz, H., Ashburner, M., Ish-Horowicz, D., and Holden, J. J.,** Genetic characterization of the 87C region of the third chromosome of *Drosophila melanogaster, Genetics,* 93, 917, 1979.
146. **Hilliker, A. J., Clark, S. H., Chovnick, A., and Gelbart, W. M.,** Cytogenetic analysis of the chromosomal region immediately adjacent to the *rosy* locus in *Drosophila melanogaster, Genetics,* 95, 95, 1980.
147. **McCarron, M., O'Donnell, J., Chovnick, A., Bhullar, B. S., Hewitt, J., and Candido, E. P. M.,** Organization of the *rosy* locus in *Drosophila melanogaster:* further evidence in support of a cis-acting control element adjacent to the xanthine dehydrogenase structural element, *Genetics,* 91, 275, 1979.
148. **Spradling, A. C. and Rubin, G. M.,** The effect of chromosomal position on the expression of the *Drosophila* xanthine dehydrogenase gene, *Cell,* 34, 47, 1983.
149. **Hall, L. M. C., Mason, P. J., and Spierer, P.,** Transcripts, genes and bands in 315,000 base-pairs of *Drosophila* DNA, *J. Mol. Biol.*, 169, 83, 1983.
150. **Bossy, B., Hall, L. M. C., and Spierer, P.,** Genetic activity along 315 kb of the *Drosophila* chromosome, *The EMBO J.*, 3, 2537, 1984.
151. **Duncan, I. W. and Kaufman, T. C.,** Cytogenetic analysis of chromosome 3 in *Drosophila melanogaster:* mapping of the proximal portion of the right arm, *Genetics,* 80, 733, 1975.
152. **Kaufman, T. C.,** Cytogenetic analysis of chromosome 3 in *Drosophila melanogaster:* isolation and characterization of four new alleles of the *proboscipedia (pb)* locus, *Genetics,* 90, 579, 1978.
153. **Kaufman, T. C., Lewis, R., and Wakimoto, B.,** Cytogenetic analysis of chromosome 3 in *Drosophila melanogaster:* the homeotic gene complex in polytene chromosome interval 84A-B, *Genetics,* 94, 115, 1980.
154. **Lewis, R. A., Kaufman, T. C., Denell, R. E., and Tallerico, P.,** Genetic analysis of the *Antennapedia* gene complex (ANT-C) and adjacent chromosomal regions of *Drosophila melanogaster.* I. Polytene chromosome segments 84B-D, *Genetics,* 95, 367, 1980.
155. **Lewis, R. A., Wakimoto, B. T., Denell, R. E., and Kaufman, T. C.,** Genetic analysis of the *Antennapedia* gene complex (ANT-C) and adjacent chromosomal regions of *Drosophila melanogaster.* II. Polytene chromosome segment 84A-84B1,2, *Genetics,* 95, 383, 1980.
156. **Denell, R. E., Hummels, K. R., Wakimoto, B. T., and Kaufman, T. C.,** Developmental studies of lethality associated with the *Antennapedia* gene complex in *Drosophila melanogaster, Dev. Biol.*, 81, 43, 1981.
157. **Wakimoto, B. T. and Kaufman, T. C.,** Analysis of larval segmentation in lethal genotypes associated with the *Antennapedia* gene complex in *Drosophila melanogaster, Dev. Biol.*, 81, 51, 1981.
158. **Garber, R. L., Kuroiwa, A., and Gehring, W. J.,** Genomic and cDNA clones of the homeotic locus *Antennapedia* in *Drosophila, The EMBO J.*, 2, 2027, 1983.
159. **Levine, M., Hafen, E., Garber, R. L., and Gehring, W. J.,** Spatial distribution of *Antennapedia* transcripts during *Drosophila* development, *The EMBO J.*, 2, 2037, 1983.
160. **Scott, M. P., Weiner, A. J., Hazelrigg, T. I., Polisky, B. A., Pirrotta, V., Scalenghe, F., and Kaufman, T. C.,** The molecular organization of the *Antennapedia* locus of *Drosophila, Cell,* 35, 763, 1983.

161. **McGinnis, W., Levine, M. S., Hafen, E., Kuroiwa, A., and Gehring, W. J.,** A conserved DNA sequence in homoeotic genes of *Drosophila Antennapedia* and *bithorax* complexes, *Nature,* 308, 428, 1984.
162. **Kuroiwa, A., Hafen, E., and Gehring, W. J.,** Cloning and transcriptional analysis of the segmentation gene *fushi tarazu* of *Drosophila, Cell,* 37, 825, 1984.
163. **Scott, M. P.,** Molecules and puzzles from the *antennapedia* homoeotic gene complex of *Drosophila, Trends in Genetics,* 1, 74, 1985.
164. **Lewis, E. B.,** Genes and gene complexes, in *Heritage from Mendel,* Brink, R. A. and Styles, E. D., Eds., University of Wisconsin Press, Madison, 1967, 17.
165. **Modolell, J., Bender, W., and Meselson, M.,** *Drosophila melanogaster* mutations suppressible by the suppressor of *Hairy-wing* are insertions of a 7.3-kilobase mobile element, *Proc. Natl. Acad. Sci. U.S.A.,* 80, 1678, 1983.
166. **Kaufman, T. C., Tasaka, S. E., and Suzuki, D. T.,** The interaction of two complex loci, *zeste* and *bithorax* in *Drosophila melanogaster, Genetics,* 75, 299, 1973.
167. **Lawrence, P. A., Johnston, P., and Struhl, G.,** Different requirements for homeotic genes in the soma and germ line of *Drosophila, Cell,* 35, 27, 1983.
168. **Wedeen, C., Harding, K., and Levine, M.,** Spatial regulation of *Antennapedia* and *bithorax* gene expression by *Polycomb* locus in *Drosophila, Cell,* 44, 739, 1986.
169. **Lawrence, P. A. and Morata, G.,** The elements of the *bithorax* complex, *Cell,* 35, 595, 1983.
170. **Akam, M.,** Decoding the *Drosophila* complexes, *Trends in Biochemical Sciences,* May, 173, 1983.
171. **North, G.,** Cloning the genes that specify fruit flies, *Nature, News and Views,* 303, 134, 1983.
172. **Bender, W., Akam, M., Karch, F., Beachy, P. A., Peifer, M., Spierer, P., Lewis, E. B., and Hogness, D. S.,** Molecular genetics of the *bithorax* complex in *Drosophila melanogaster, Science,* 221, 23, 1983.
173. **Marx, J.,** Genes that guide fruit fly development, *Science, Research News,* 223, 1223, 1984.
174. **North, G.,** How to make a fruitfly, *Nature, News and Views,* 311, 214, 1984.
175. **Ingham, P.,** The regulation of the *bithorax* complex, *Trends in Biochemical Sciences,* April, 112, 1985.
176. **Anderson, K. V., Jūrgens, G., and Nūsslein-Volhard, C.,** Establishment of dorsal-ventral polarity in the *Drosophila* embryo: genetic studies on the role of the *Toll* gene product, *Cell,* 42, 779, 1985.
177. **Ingham, P. W.,** A gene that regulates the *bithorax* complex differentially in larval and adult cells of *Drosophila, Cell,* 37, 815, 1984.
176. **Haenlin, M., Steller, H., Pirrotta, V., and Mohier, E.,** A 43 kilobase cosmid P transposon rescues the *fs(1)K10* morphogenetic locus and three adjacent *Drosophila* developmental mutants, *Cell,* 40, 827, 1985.
179. **Steward, R., McNally, F. J., and Schedl, P.,** Isolation of the *dorsal* locus of *Drosophila, Nature,* 311, 262, 1884.
180. **Kornberg, T.,** *Engrailed:* a gene controlling compartment and segment formation in *Drosophila, Proc. Natl. Acad. Sci. U.S.A.,* 78, 1095, 1981.
181. **Kornberg, T.,** Compartments in the abdomen of *Drosophila* and the role of the *engrailed* locus, *Dev. Biol.,* 86, 363, 1981.
182. **Fjose, A., McGinnis, W. J., and Gehring, W. J.,** Isolation of a homoeo box-containing gene from the *engrailed* region of *Drosophila* and the spatial distribution of its transcripts, *Nature,* 313, 284, 1985.
183. **Poole, S. J., Kauvar, L. M., Drees, B., and Kornberg, T.,** The *engrailed* locus of *Drosophila:* structural analysis of an embryonic transcipt, *Cell,* 40, 37, 1985.
184. **Spencer, F. A., Hoffmann, F. M., and Gelbart, W. M.,** *Decapentaplegic:* a gene complex affecting morphogenesis in *Drosophila melanogaster, Cell,* 28, 451, 1982.
185. **Rubin, G. M.,** Summary, *Cold Spring Harbor Symp. Quant. Biol.,* 50, 905, 1986.
186. **Hochman, B., Gloor, H., and Green, M. M.,** Analysis of chromosome 4 in *Drosophila melanogaster.* I. Spontaneous and X-ray-induced lethals, *Genetica,* 35, 109, 1964.
187. **Hochman, B.,** A note on the salivary chromosome 4 in *D. melanogaster, Drosophila Inf. Serv.,* 40, 67, 1965.
188. **Hochman, B.,** EMS and ICR-100 induced chromosome 4 lethals in *D. melanogaster, Drosophila Inf. Serv.,* 42, 59, 1967.
189. **Hochman, B.,** Analysis of chromosome 4 in *Drosophila melanogaster.* II. Ethyl methanesulfonate induced lethals, *Genetics,* 67, 235, 1971.
190. **Hochman, B.,** Analysis of a whole chromosome in *Drosophila, Cold Spring Harbor Symp. Quant. Biol.,* 38, 581, 1974.
191. **Bonner, J. J. and Pardue, M. L.,** Ecdysone-stimulated RNA synthesis in salivary glands of *Drosophila melanogaster:* assay by in situ hybridization, *Cell,* 12, 219, 1977.
192. **Postlethwait, J. H. and Jowett, T.,** Genetic analysis of the hormonally regulated yolk polypeptide genes in *D. melanogaster, Cell,* 20, 671, 1980.
193. **Spradling, A. C. and Mahowald, A. P.,** Identification and genetic localization of mRNAs from ovarian follicle cells of *Drosophila melanogaster, Cell,* 16, 589, 1979.
194. **Spradling, A. C.,** The organization and amplification of two chromosomal domains containing *Drosophila* chorion genes, *Cell,* 27, 193, 1981.

195. **Sanchez, F., Natzle, J. E., Cleveland, D. W., Kirschner, M. W., and McCarthy, B. J.,** A dispersed multigene family encoding tubulin in *Drosophila melanogaster, Cell,* 22, 845, 1980.
196. **Bialojan, S., Falkenburg, D., and Renkawitz-Pohl, R.,** Characterization and developmental expression of B tubulin genes in *Drosophila melanogaster, The EMBO J.,* 3, 2543, 1984.
197. **Fyrberg, E. A., Kindle, K. L., and Davidson, N.,** The actin genes of *Drosophila:* a dispersed multigene family, *Cell,* 19, 365, 1980.
198. **Tobin, S. L., Zulauf, E., Sanchez, F., Craig, E. A., and McCarthy, B. J.,** Multiple actin-related sequences in the *Drosophila melanogaster* genome, *Cell,* 19, 121, 1980.
199. **Zulauf, E., Sanchez, F., Tobin, S. L., Rdest, U., and McCarthy, B. J.,** Developmental expression of a *Drosophila* actin gene encoding actin I, *Nature,* 292, 556, 1981.
200. **Sodja, A., Rizki, R. M., Rizki, T. M., and Zafar, R. S.,** Overlapping deficiencies refine the map position of the sex-linked actin gene of *Drosophila melanogaster, Chromosoma,* 86, 293, 1982.
201. **Korge, G.,** Gene activities in larval salivary glands of insects, *Verh. Dtsch. Zool. Ges.,* 1980, 94, 1980.
202. **Velissariou, V. and Ashburner, M.,** The secretory proteins of the larval salivary gland of *Drosophila melanogsater.* Cytogenetic correlation of a protein and a puff, *Chromosoma,* 77, 13, 1980.
203. **Velissariou, V. and Ashburner, M.,** Cytogenetic and genetic mapping of a salivary gland secretion protein in *Drosophila melanogaster, Chromosoma,* 84, 173, 1981.
204. **Korge, G.,** Genetic analysis of the larval secretion gne *Sgs-4* and its regulatory chromosome sites in *Drosophila melanogaster, Chromosoma,* 84, 373, 1981.
205. **McGinnis, W., Shermoen, A. W., and Beckendorf, S. K.,** A transposable element inserted just 5′ to a *Drosophila* glue protein gene alters gene expression and chromatin structure, *Cell,* 34, 75, 1983.
206. **Crowley, T. E., Mathers, P. H., and Meyerowitz, E. M.,** A trans-acting regulatory product necessary for expression of the *Drosophila melanogaster* 68C glue gene cluster, *Cell,* 39, 149, 1984.
207. **Peterson, N. S., Moller, G., and Mitchell, H. K.,** Genetic mapping of the coding regions for three heat-shock proteins in *Drosophila melanogaster, Genetics,* 92, 891, 1979.
208. **Lis, J. T., Prestidge, L., and Hogness, D. S.,** A novel arrangement of tandemly repeated genes at a major heat shock site in *Drosophila melanogaster, Cell,* 14, 901, 1978.
209. **Craig, E. A., McCarthy, B. J., and Wadsworth, S. C.,** Sequence organization of two recombinant plasmids containing genes for the major heat shock-induced protein of *Drosophila melanogaster, Cell,* 16, 575, 1979.
210. **Ish-Horowicz, D., Pinchin, S. M., Gausz, J., Gyurkovics, H., Bencze, G., Goldschmidt-Clermont, M., and Holden, J. J.** Deletion mapping of two *D. melanogaster* loci that code for the 70,000 dalton heat-induced protein, *Cell,* 17, 565, 1979.
211. **Scalenghe, F. and Ritossa, F.,** The puff inducible in region 93D is responsible for the synthesis of the major "heat shock" polypeptide in *Drosophila melanogaster, Chromosoma,* 63, 317, 1977.
212. **Walldorf, U., Richter, S., Ryseck, R.-P., Steller, H., Edström, J. E., Bautz, E. K. F., and Hovemann, B.,** Cloning of heat-shock locus 93D from *Drosophila melanogaster, The EMBO J.,* 3, 2499, 1984.
213. **Grond, C. J., Peters, F. P. A. M. N., Derksen, J., and van der Ploeg, M.,** Identification of the heat shock band 2-48B of *Drosophila hydei* and determination of its haploid DNA content, *Eur. J. Cell Biol.,* 31, 150, 1983.
214. **Steffensen, D. M. and Wimber, D. E.,** Localization of tRNA genes in the salivary chromosomes of *Drosophila* by RNA:DNA hybridization, *Genetics,* 69, 163, 1971.
215. **Grigliatti, T. A., White, B. N., Tener, G. M., Kaufman, T. C., Holden, J. J., and Suzuki, D. T.,** Studies on the transfer RNA genes of *Drosophila, Cold Spring Harbor Symp. Quant. Biol.,* 38, 461, 1974.
216. **Grigliatti, T. A., White, B. N., Tener, G. M., Kaufman, T. C., and Suzuki, D. T.,** The localization of transfer RNA_5^{Lys} genes in *Drosophila melanogaster, Proc. Natl. Acad. Sci. U.S.A.,* 71, 3527, 1974.
217. **Hayashi, S., Gillam, I. C., Grigliatti, T. A., and Tener, G. M.,** Localization of tRNA genes of *Drosophila melanogsater* by in situ hybridization, *Chromosoma,* 86, 279, 1982.
218. **Dunn, R., Hayashi, S., Gillam, I. C., Delaney, A. D., Tener, G. M., Grigliatti, T. A., Kaufman, T. C., and Suzuki, D. T.,** Genes coding for valine transfer ribonucleic acid-3b in *Drosophila melanogaster, J. Mol. Biol.,* 128, 277, 1979.
219. **Hayashi, S., Gillam, I. C., Delaney, A. D., Dunn, R., Tener, G. M., Grigliatti, T. A., and Suzuki, D.,** Hybridization of tRNAs of *Drosophila melanogaster* to polytene chromosomes, *Chromosoma,* 76, 65, 1984.
220. **Dudler, R., Schmidt, T., Bienz, M., and Kubli, E.,** The genes coding for $tRNA^{tyr}$ of *Drosophila melanogaster:* localization and determination of the gene numbers,*Chromosoma,* 84, 49, 1981.
221. **Myslinski, E., Branlant, C., Wieben, E. D., and Pederson, T.,** The small nuclear RNAs of *Drosophila, J. Mol. Biol.,* 180, 927, 1984.
222. **Ritossa, F. M. and Spiegelman, S.,** Localization of DNA complementary to ribosomal RNA in the nucleolus organizer region of *Drosophila melanogaster, Proc. Natl. Acad. Sci. U.S.A.,* 53, 737, 1965.
223. **Pardue, M. L., Gerbi, S. A., Eckhardt, R. A., and Gall, J. G.,** Cytological localization of DNA complementary to ribosomal RNA in polytene chromosomes of *Diptera, Chromosoma,* 29, 268, 1970.

224. **Glover, D. M., White, R. L., Finnegan, D. J., and Hogness, D. S.,** Characterization of six cloned DNAs from *Drosophila melanogaster,* including one that contains the genes for rRNA, *Cell,* 5, 149, 1975.
225. **Glover, D. M. and Hogness, D. S.,** A novel arrangement of the 18S and 28S sequences in a repeating unit of *Drosophila melanogaster* rDNA, *Cell,* 10, 167, 1977.
226. **Endow, S. A.,** On ribosomal gene compensation in *Drosophila, Cell,* 22, 149, 1980.
227. **Wellauer, P. K. and Dawid, I. B.,** the structural organization of ribosomal DNA in *Drosophila melanogaster, Cell,* 10, 193, 1977.
228. **Pellegrini, M., Manning, J., and Davidson, N.,** Sequence arrangement of the rDNA of *Drosophila melanogster, Cell,* 10, 213, 1977.
229. **Dawid, I. B., Wellauer, P. K., and Long, E. O.,** Ribosomal DNA in *Drosophila melanogaster.* I. Isolation and characterization of cloned fragments, *J. Mol. Biol.,* 126, 749, 1978.
230. **Wellauer, P. K. and Dawid, I. B.,** Ribosomal DNA in *Drosophila melanogaster.* II. Heteroduplex mapping of cloned and uncloned rDNA, *J. Mol. Biol.,* 126, 769, 1978.
231. **Wellauer, P. K., Dawid, I. B., and Tartof, K. D.,** X and Y chromosomal ribosomal DNA of *Drosophila:* comparison of spacers and insertions, *Cell,* 14, 269, 1978.
232. **Endow, S. A. and Glover, D. M.,** Differential replication of ribosomal gene repeats in polytene nuclei of *Drosophila, Cell,* 17, 597, 1979.
233. **Barnett, T. and Rae, P. M. M.,** A 9.6 kb intervening sequence in *D. virilis* rDNA, and sequence homology in rDNA interruptions of diverse species of *Drosophila* and other *Diptera, Cell,* 16, 763, 1979.
234. **Tartof, K. D.,** Evolution of transcribed and spacer sequences in the ribosomal RNA genes of *Drosophila, Cell,* 17, 607, 1979.
235. **Rae, P. M. M., Barnett, T., and Murtif, V. L.,** Nontranscribed spacers in *Drosophila* ribosomal DNA, *Chromosoma,* 82, 637, 1981.
236. **Kunz, W., Petersen, G., Renkawitz-Pohl, R., Glätzer, K. H., and Schäfer, M.,** Distribution of spacer length classes and the intervening sequence among different nucleolus organizers in *Drosophila hydei, Chromosoma,* 83, 145, 1981.
237. **Hennig, W., Vogt, P., Jacob, G., and Siegmund, I.,** Nucleolus organizer regions in *Drosophila* species of the *repleta* group, *Chromosoma,* 87, 279, 1982.
238. **Ising, G.,** A recessive lethal in chromosome 2, which in single dose has an effect on the eye colour of white animals, *Drosophila Inf. Serv.,* 39, 84, 1964.
239. **Ising, G. and Ramel, C.,** A white-suppressor behaving as an episome in *Drosophila melanogster, Genetics (Suppl.),* 74, 123, 1971.
240. **Ising, G. and Block, K.,** Derivation-dependent distribution of insertion sites for a *Drosophila* transposon, *Cold Spring Harbor Symp. Quant. Biol.,* 45, 527, 1981.
241. **Finnegan, D. J. and Fawcett, D. H.,** Transposable elements in *Drosophila melanogaster,* in the *Oxford Surveys on Eukaryotic Genes,* Vol. 3, Oxford University Press, Oxford, 1986, 1.
242. **Young, M. W. and Schwartz, H. E.,** Nomadic gene families in *Drosophila, Cold Spring Harbor Symp. Quant. Biol.,* 45, 629, 1981.
243. **Tchurikov, N. A., Ilyin, Y. V., Skryabin, K. G., Ananiev, E. V., Bayev, A. A., Jr., Krayev, A. S., Zelentsova, E. S., Kulguskin, V. V., Lyubomirskaya, Nv, and Georgeiv, G. P.,** General properties of mobile dispersed genetic elements in *Drosophila melanogaster, Cold Spring Harbor Symp. Quant. Biol.,* 45, 655, 1981.
244. **Gvozdev, V. A., Belyaeva, E. S., Ilyin, Y. V., Amosova, I. S., and Kaidanov, L. Z.,** Selection and transposition of mobile dispersed genes in *Drosophila melanogaster, Cold Spring Harbor Symp. Quant. Biol.,* 45, 673, 1981.
245. **Rasmuson, B., Westerberg, B. M., Rasmuson, Å., Gvozdev, V. A., Belyaeva, E. S., and Ilyin, Y. V.,** Transpositions, mutable genes, and the dispersed gene family *Dm 225* in *Drosophila melanogaster, Cold Spring Harbor Symp. Quant. Biol.,* 45, 545, 1981.
246. **Belyaeva, E. Sp., Ananiev, E. V., and Gvozdev, V. A.,** Distribution of mobile dispersed genes *(mdg-1* and *mdg-3)* in the chromosomes of *Drosophila melanogaster, Chromosoma,* 90, 16, 1984.
247. **Ananiev, E. V., Barsky, V. E., Ilyin, Yu. V., and Ryzic, M. V.,** The arrangement of transposable elements in the polytene chromosomes of *Drosophila melanogaster, Chromosoma,* 90, 366, 1984.
248. **Ananiev, E. V. and Barsky, V. E.,** Electron Microscopic Map of the Polytene Chromosomes in Salivary Glands of *Drosophila (D. melanogaster),* (in Russian), Nauka, Moscow, 1984.
249. **Merriam, J.,** Cloned DNA by chromosome location, *Drosophila Inf. Serv.,* 61, 9, 1985.
250. **Merriam, J., Smalley, S. L., Merriam, A., and Dawson, B.,** The molecular genome of *Drosophila melanogaster.* Catalogs of cloned DNA, breakpoints and transformed inserts by chromosome location, *Drosophila Inf. Serv.,* 63, 173, 1986.

Chapter 18

CHROMOSOME MAPS AND THE NEW TRENDS IN *DROSOPHILA* GENETICS: CONCLUDING REMARKS

Light and electron microscopic studies on the structural organization of salivary gland chromosomes have proven the polytene nature of these chromosomes. According to the current knowledge, the DNA of individual chromatids is continuous throughout the chromosomes. Most of the DNA of euchromatic arms is located in the chromomeric loops, the size of them ranging from about 3 to about 100 kb per chromomere. Parallel chromomeres of homologous chromatids form a single disk-like chromomere band per homolog or a number of smaller, more separate and probably toroidal groups (dotted bands). Many of the bands have their own characteristic appearance.

Aside from the variation of the polyteny degree that appears between different species of *Drosophila* and between different tissues, there is also local variation of polyteny in individual chromosomes. Between the euchromatic and heterochromatic regions this variation is evident, but it may also exist in regions composed of extremely heavy bands, due to their late or incomplete replication.

The number bands (and interphase chromomeres) in the salivary gland chromosomes of the different species of *Drosophila* may be essentially similar although their mapping has not yet been carried out with equal accuracy. The divergence found in the structural organization of polytenized interphase chromosomes derived from other tissues may represent mainly differential groupings of chromomeres. At least in part, the differential appearance of chromosomes seems to be caused by a lower or higher degree of polyteny in those cells, if compared with the chromosomes of salivary gland cells. The structural divergence is apparently pronounced in squashed chromosomes by differential stretching and unfolding of chromomeres. Both the polyteny degree and the grouping and bundling of chromomere units are obviously typical to each polytenized tissue.

An abundance of studies have also been published concerning both the natural and induced puffing activity in polytene chromosomes. An existence of numerous and site-specific puffs has been documented from different stages of larval development. It may be concluded that the principles of structural and functional organization of polytene chromosomes are known although many details and the molecular basis of them need to be solved.

The polytenized interphase chromosomes occurring in the larval salivary gland cells of *Drosophila* have been used for mapping for already more than 50 years. Thousands of relatively constant structural landmarks are included in the banding pattern of polytene chromosomes. The bands and regions composed by chromomeres having a typical axial length and arrangement can be recognized and identified although inverted or translocated into a new chromosomal position. In most of the species of *Drosophila* the salivary gland chromosome maps have been compiled to serve phylogenetic studies recording the chromosomal rearrangements taken place during the evolution and speciation. By comparison of band sequences in the salivary gland chromosomes of related species, in many cases, the phylogenetic tree can be reconstructed by following the order and type of rearrangements occurred in the chromosomes. For evolutionary studies, usually, the description of band sequences at the division or subdivision level is already informative. The drawn or photographed maps of salivary gland chromosomes are available of more than 100 species of *Drosophila*.

Because of the most advanced genetic mapping and characterization of *Drosophila melanogaster*, the polytene chromosomes of this species have also been extensively used for localization of smaller chromosomal changes and of individual genes. For the cytogenetic allocation of genes, certainly, more detailed maps, recording all the chromomere bands of the polytene chromosomes, are required. This was already realized in the early phases of mapping of the salivary gland chromosomes of *D. melanogaster*.

The DNA content of an average size of gene is only a fraction of the DNA content of an average-size chromomere. According to the electron microscopic analyses of the banding pattern, the average thickness of bands in the salivary gland chromosomes of *D. melanogaster* is about 100 nm (=0.1 μm), although about 480 Bridges' doublets are interpreted as single bands. Since most of the bands are even narrower, the accurate cytogenetic mapping of genes has to be carried out at ultimate resolution limits of the light microscope. This certainly makes great demands to the preparation techniques and equipment used in the microscopic examination, and also to the preciseness and reliability of reference maps.

Many of the new electron microscopic methods offer a higher mapping resolution than any light microscopic method, and thus they also offer possibilities to make more accurate measurements of the dimensions of bands and chromomeres in the polytene chromosomes. Since the thickness of bands can be measured starting from about 20 nm, a large variety of bands can be classified according to their average axial length determined on the basis of electron micrographs. By using the thickness values of the bands, the approximate DNA contents of chromatids can be estimated per chromomere units, subdivisions and divisions.

Thus the electron microscopic revision of the reference maps of the salivary gland chromosomes of *D. melanogaster* makes it possible to approximate DNA distances between regions, bands, and genes. This may help the geneticists starting molecular studies on certain regions to plan chromosome jumps and walks. Accordingly, the electron microscopic verification and revision of the light microscopic reference maps allows more accurate localization of deficiencies and borders of larger rearrangements, as well as of the sites of transcription and hybridization along the salivary gland chromosomes.

During the last few years many genes controlling important metabolic and developmental processes in *Drosophila* have been cloned and analyzed. Particularly, the genes involved in the early phases of embryogenesis have been subjected to extensive studies. Some of the findings concerning the molecular organization of developmental genes in *Drosophila* have had also wider influence. For instance, the homeo box domain, found to be characteristic to many genes controlling the segmentation stage in early embryonic development in *Drosophila*, has been detected in many other organisms. Particularly, the Antennapedia-type of homeo box domain seems to be widely distributed and share remarkable homology to similar domains found in the genomes of higher metazoans, like Annelids, Arthropods, Chordates and Ascidians. The finding of the homeo box from also Echinoderms points to an even older origin of this highly conserved domain. It also points to a more general role of this domain in the coordination of the interactions of developmental genes.

The polytene chromosomes have rendered it possible to accurately map developmental and other genes. The polytene chromosomes have also made it possible to isolate DNA from the given bands or regions for direct analysis or for micro cloning. This essentially helps the determination of the origin of DNA and, practically, allows to start molecular studies from any region identifiable with the reference maps. Thus it seems that polytene chromosomes are going to have an increasing importance as a source of chromosomal DNA, as well as a concrete gene map. In other words, the polytene chromosomes, only, offer a resolution that is satisfactory to create techniques for direct isolation of genes and to localize gene DNA obtained by using other methods.

INDEX

E

F

G

H

I

K

L

M

N

O

P